Fatigue analysis of random loadings

Denis Benasciutti

Fatigue analysis of random loadings

A frequency-domain approach

LAP LAMBERT Academic Publishing

Imprint

Any brand names and product names mentioned in this book are subject to trademark, brand or patent protection and are trademarks or registered trademarks of their respective holders. The use of brand names, product names, common names, trade names, product descriptions etc. even without a particular marking in this work is in no way to be construed to mean that such names may be regarded as unrestricted in respect of trademark and brand protection legislation and could thus be used by anyone.

Cover image: www.ingimage.com

Publisher:
LAP LAMBERT Academic Publishing
is a trademark of
International Book Market Service Ltd., member of OmniScriptum Publishing Group
17 Meldrum Street, Beau Bassin 71504, Mauritius

ISBN: 978-3-659-12370-2

Zugl. / Approved by: Ferrara, University of Ferrara, Diss., 2005

TABLE OF CONTENTS

Chapter 1

INTRODUCTION

It is a common opinion that most of the failures observed in real structures and mechanical components are due to fatigue. In the design of real systems subjected to environmental loadings both fatigue strength and dynamic properties of external loads are
important. However, the fatigue phenomenon is a quite complicate process, which is affected by large scatter.

Material fatigue strength depends on the amplitudes of loading cycles; experimental
tests using constant amplitude cycles are generally performed on standard specimens to
estimate a S-N curve, in which the number of cycles to failure is expressed as a function
of the amplitude level.

Fatigue damage accumulation can be explained in terms of initiation and growth of
small cracks into the metal; crack propagation progressively reduces the capability of
specimens to carry the applied external load, so that they finally break. The crack nucleation that will eventually cause failure depends on local conditions, such as microscopic inhomogeneities, grain structure or microscale defects. This dependency is responsible of the large scatter generally observed in the fatigue strength.

The external loads measured in real systems are often quite irregular, see Figure 1.1.
Examples are the loads generated by wind and/or waves (in wind turbines, ship details,
off-shore platforms), or those produced on vehicles by the irregularity of road profile.
This randomness, of course, increases the overall uncertainty of fatigue life prediction
and should be adequately accounted for.

In order to determine the fatigue damage that an irregular load causes to the material,
counting methods and damage accumulation models are generally adopted. Amongst all
counting methods, the rainflow count is widely regarded as the best counting procedure,
while the Palmgren-Miner linear damage rule is usually adopted for its simplicity.

A possible approach is to model an irregular load as a stationary random process and
to use frequency-domain methods (which refer to the spectral density of the process) to
estimate the statistical distribution of rainflow cycle, the fatigue damage and the system
service life. Interesting results have also been obtained under the Markov chain ap-

proximation for the sequence of extremes [Rychlik 1987, Rychlik 1989, Bishop and Sherrat 1990, Frendhal and Rychlik 1993, Bishop 1994].

The hypothesis of Gaussian process is generally used by most of the methods available in the literature, since it is a simplifying assumption that allows deriving explicit (exact or approximate) formulas for cycle distribution and fatigue damage (e.g. the Rayleigh amplitude distribution for a narrow-band process).

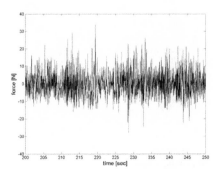

Figure 1.1: Force measured on a bicycle handlebar.

However, in most applications concerning real structural details subjected to random excitations, the loading response may not satisfy the Gaussian assumption; in fact, the external excitation may be itself non-Gaussian (e.g. wave or wind loads), or the system may be non-linear, so that a Gaussian input is converted into a non-Gaussian internal response, or both.

The literature concerning the fatigue analysis of non-Gaussian loadings highlights that any deviation from the Gaussian hypothesis may be responsible of a considerable increase of the overall rate of fatigue damage accumulation [Sarkani et al. 1994, Jha and Winterstein 2000]. Then, spectral methods valid for Gaussian processes may greatly underestimate the fatigue damage rate (thus overestimating the system service life) in non-Gaussian loadings. Then, application of spectral methods to non-Gaussian loadings should account for the deviation from the Gaussian hypothesis.

Furthermore, all methods existing in literature are usually restricted to random loadings assumed as stationary, even if in many applications this hypothesis may be too restrictive. For example, measured stress responses may be non-stationary, as those resulting from different wind conditions, various sea states, as well as road sequences with different surface profile characteristics. In the case of piece-wise stationary loadings (i.e. switching processes), interesting results have been achieved under the Markov approximation [Johannesson 1998, Johannesson 1999, Johannesson et al. 1995, Benasciutti and Tovo 2010].

Moreover, it is worth noting that mechanical components with complex geometries could be subjected to a multiaxial state of stress. Consequently, the fatigue analysis scheme obviously becomes more complex. Frequency-domain approaches for multiaxial loadings case can be found in the literature [Pitoiset et al. 1998, Pitoiset and Preu-

2

mont 2000, Pitoiset 2001, Pitoiset et al. 2001, Benasciutti and Cristofori 2008, Cristofori et al. 2011a]. Nevertheless, the full understanding of the process of fatigue damage accumulation under multiaxial loading components and the definition of a multiaxial cycle counting method [Langlais et al. 2003], as well as the identification of correct multiaxial fatigue criteria, are still investigated.

This book concentrates on the fatigue analysis of both Gaussian and non-Gaussian uniaxial stationary random loadings. Characterisation of random loadings in both time- and frequency-domain (e.g. definition of spectral parameters) is given, as well as the characterisation of the distribution of rainflow cycles (i.e. definition of the joint distribution $h(u, v)$ of counted cycles and its properties) and its related fatigue damage. General relations amongst bandwidth parameters are given in Appendix A.

Two original approaches (i.e. the TB method and the empirical-$\alpha_{0.75}$ method), as well as several other methods valid for Gaussian processes (e.g. Wirsching-Light, Dirlik and Zhao-Baker method), are presented. Numerical simulations are used to compare all methods and to investigate some correlations existing between fatigue damage and spectral parameters.

Bimodal Gaussian random processes, important in wind, off-shore and automotive applications, are also treated. Some spectral methods, specific for bimodal processes (e.g. single-moment, Jiao-Moan, Sakai-Okamura, Fu-Cebon method), are discussed. A modification of the Fu-Cebon approach is also proposed. Bimodal spectral densities are used in numerical simulations to find the range of applicability of each method.

Finally, a simple application, concerning a linear two-degrees-of-freedom model moving on an irregular profile (used to approximate a car-quarter on a road), is proposed, which aims to underline all capacities and potentialities of frequency-domain analysis in the design and sensitivity analysis of real components even at early stages.

The fatigue analysis of non-Gaussian random loadings is then discussed in its general theoretical aspects (e.g. definition of the transformed Gaussian model) and some general properties of the rainflow cycle distribution are given. The TB method is further extended to the non-Gaussian case: it is described in its main theoretical aspects and it is then applied to real data taken from a Mountain-bike and from an automotive component.

The book concludes by comparing the experimental lifetime data taken from the literature and referring to tests on a welded joint under non-Gaussian random loadings with theoretical results given by different spectral methods.

All subjects presented in this book and their connection to fatigue analysis of random loadings are summarised in Figure 1.2.

For the definition of counting methods described in Chapter 2, we mainly refer to the ASTM standard [ASTM 1985] and to Dowling [Dowling 1972]. The theory of random processes, as well as the definition of bandwidth parameters, is presented by referring to the book by Lutes and Sarkani [Lutes and Sarkani 2004]. For what concerns the non-Gaussian loadings, different definitions of parametric models for the transformation linking the non-Gaussian to the underlying Gaussian domain are found in [Winterstein 1988, Sarkani et al., Ochi and Ahn 1994]. A non-parametric definition is that proposed in [Rychlik et al. 1997].

All numerical simulations are performed using MATLAB; the WAFO (Wave Analysis for Fatigue and Oceanography) toolbox[*] is also used. New routines are written, in particular for the fatigue analysis of non-Gaussian loadings.

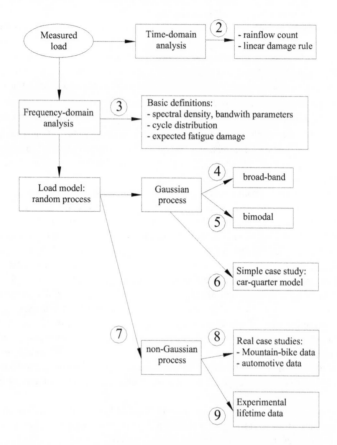

Figure 1.2: Overview of the book. Numbers indicate in which chapter the information can be found.

This book is mainly devoted to the analysis of uniaxial stationary loadings. The problem of stress multiaxiality is partly addressed in Chapter 8, while more recent developments concerning multiaxial spectral methods can be found, for example, in [Susmel et al. 2009, Cristofori et al. 2011a]. The problem of load non-stationarity, instead, is

---

[*] The toolbox can be downloaded free of charge from the website of the Centre of Mathematical Statistics, Lund University: http://www.maths.lth.se/matstat/wafo/.

not investigated in the book; interesting results concerning the frequency-domain analysis of "switching loadings" can be found in [Benasciutti and Tovo 2007b, 2010].

1.1. FATIGUE DAMAGE UNDER RANDOM LOADINGS

The fatigue process occurring in real components under random loadings is a quite complex phenomenon.

It is generally agreed that fatigue damage only depends on the sequence of maxima and minima (i.e. the sequence of turning points) in a given load, i.e. damage accumulation is rate-independent. However, it is no obvious how to define fatigue cycles in a variable amplitude load, hence we use appropriate counting methods. Moreover, the process of damage accumulation requires the definition of suitable damage accumulation rules. Furthermore, there are also many uncertainties both in fatigue strength determination and in load characterisation.

It is very important to distinguish in the fatigue assessment procedure all sources of uncertainties. Casualties intrinsic into the fatigue process, as those regarding material properties, should be distinguished from uncertainties deriving from a lack of knowledge of the true parameters involved, or from the randomness of the external load, as well as randomness due to incorrect modelling.

According to Svensson [Svensson 1997], all uncertainties in the fatigue phenomenon could be divided into five items:

1. material properties;
2. structural properties;
3. load variation;
4. parameter estimation;
5. model errors

It is necessary to note that causes listed above are quite different. In fact, the first three items define the intrinsic random nature of the fatigue process, while the other two only follows from the lack of knowledge of the true parameters involved (usually due to the limited amount of experimental data) or of the fatigue damage mechanism under variable amplitude loading (see [Tovo 2001]).

Material properties

The fatigue strength of a material or structural detail primarily depends on the amplitudes of loading cycles. In experimental tests, the resistance to fatigue is usually expressed as the number of cycles to failure under constant amplitude loading cycles.

Larger amplitudes are observed to give shorter fatigue lives; in addition, many materials exhibit a fatigue limit (or endurance limit) s_∞, i.e. an amplitude level below which fatigue life can be infinite, although the existence of a fatigue limit is under debate [Sonsino 2007a] . Mathematically, the fatigue resistance is usually expressed as a power-law function (called S-N or Wöhler curve):

$$N = \begin{cases} 0 & s_{\max} \geq S_{u} \\ C s^{-k} & s > s_{\infty} \ , s_{\max} < S_{u} \\ \infty & s \leq s_{\infty} \end{cases} \qquad (1.1)$$

where N is the number of cycles to failure at amplitude s, $s_{\max}$ is the maximum of the loading cycle and S_{u} is the ultimate tensile strength. The material parameters are k (the S-N slope) and C (the fatigue strength). The exponent k typically ranges from 2 to 4 for notched components.

Note that the above equation accounts for both the fatigue limit and the static strength; in all calculations performed throughout the book, we shall refer instead to a simplified S-N curve:

$$s^{k} N = C \qquad (1.2)$$

It is well known that components of the same kind show a large scatter in their fatigue life, even if they experience exactly the same nominal loading conditions. Therefore, some or all of the parameters in the S-N curve should be modelled as random variables (see [Lindgren and Rychlik 1987, Svensson 1997]).

Structural properties

Experimental laboratory tests to determine the material fatigue strength are usually performed on standard specimens under highly controlled conditions.

In engineering design problems, the properties of complex structural details must be considered. Usually, the fatigue strength of real components is derived from that of the material by accounting for the geometry and the surface finishing. However, it can happen that the technological process during manufacturing (e.g. different welding technologies) influences the fatigue strength. In some circumstances, it is also important to account for a corrosive environment.

Load variation

The simplest kind of loading condition is constant amplitude loading used in fatigue strength determination.

In some circumstances, there can be block loadings formed by sequences of cycles having different amplitudes. The common approach is to adopt a damage rule to account for different damage produced by cycles with different amplitudes.

The Palmgren-Miner damage accumulation rule affirms that damage increases linearly during cycle repetitions. In a constant amplitude test, each cycle with amplitude s then uses a fraction $1/N$ of the total fatigue life. The total fatigue damage is then computed as:

$$D = \sum \frac{n_i}{N_i} \qquad (1.3)$$

where n_i is the number of cycles with amplitude s_i. Fatigue failure is assumed to occur when damage reaches a critical value D_{cr}, i.e. $D = D_{cr}$. Critical damage value D_{cr} is often assumed unity, although this value is shown to be on the unsafe side by experimental tests on un-welded and welded components [Sonsino 2007b].

The advantage of the linear law is its simplicity, even if the damage measure is not related to a direct physical quantity (such as crack length) and it ignores load interaction (sequence) effects.

The Palmgren-Miner damage rule is adopted in the fatigue analysis of service loadings. Fatigue in service is characterised by a load variation which shows a random behaviour and increases the overall uncertainty.

In particular, in service the load variation is not known beforehand, and thus the main problem is the statistical characterisation of the main properties of the load.

Parameter estimation

Most of the models used in fatigue analysis are parametric and the parameters are often estimated by fitting the model to experimental data; an example is given by the S-N curve. If parameters in the model are assumed as random variables, the uncertainty in parameter estimation reflects into the uncertainty of the fatigue damage estimation.

Model errors

The crack nucleation depends on many local material conditions and no physical model exists that can account for all these dependencies.

In addition, even if there were physical models for the crack development, there would be no possibility from an engineering point of view to estimate component fatigue life by examining all microstructural features.

Therefore, the complexity of fatigue process makes it necessary to use simplified models in order to predict fatigue life.

As an example, the simple S-N curve describes the average fatigue life under constant amplitude loadings. However, the life in different tests is often observed to vary around this average life, and this variation can be evaluated by randomising the material parameters k and C.

1.2. OVERVIEW OF THE BOOK

The book is focused on the fatigue analysis of irregular loadings, as those measured in real mechanical components and structures under service conditions, with particular emphasis on a frequency-domain approach.

The work is organised in two parts, which reflect the two possible alternative methodologies usually adopted in the fatigue assessment of real components, i.e. time- and frequency-domain methods. Time-domain methods are based on cycle counting algorithms and damage accumulation rules, and are applied to deterministic loads (e.g. loads measured experimentally on real systems or prototypes). Frequency-domain methods are developed on the model of random process for the irregular load: stationary random processes, described by a spectral density function, are analysed. According to shape of their spectral density, random processes are narrow-band when their spectral density extends over a restrict range of frequencies, and broad-band (or wide-band) when their spectral density possess a wider frequency content.

Spectral moments and bandwidth parameters are used to quantify to what extent a particular process can be considered either narrow-band or broad-band.

The book investigates Gaussian and non-Gaussian random processes. The overview of the book is shown in Figure 1.2.

1.2.1. Counting methods for deterministic loadings

In Chapter 2 all the necessary definitions important for fatigue analysis are presented, e.g. turning points, level-crossings, amplitude and mean values of fatigue cycles.

The most used counting methods for variable amplitude loads are reviewed: peak-valley, level-crossing, range, range-pairs and rainflow counting. All methods are applied to a short time history used as an example, and results from different cycle counts are provided. The Palmgren-Miner linear damage accumulation rule is also discussed.

1.2.2. Random processes, cycle distribution and fatigue damage

The concept of stationary random process is used to model an irregular loading measured in a structure. The random process is characterised in frequency-domain by a spectral density function, which is described in terms of spectral moments and bandwidth parameters. General relations involving all spectral parameters are summarised in Appendix A. Subsequently, the description of the distribution of rainflow cycles and the related fatigue damage follows. The main problem in the fatigue analysis of random loadings is to find the true shape of the probability distribution of rainflow cycles and the correlation between this distribution and the process spectral density.

1.2.3. Fatigue analysis of Gaussian loadings

In Chapter 4 we discuss the most used spectral methods valid for uniaxial Gaussian random processes: the narrow-band approximation and the Wirsching-Light damage correction formula [Wirsching and Light 1980], the Dirlik [Dirlik 1985] and the Zhao-Baker [Zhao and Baker 1992] model for the amplitude probability density of rainflow cycles, the Petrucci-Zuccarello [Petrucci and Zuccarello 2001] approximate rainflow damage. The method originally proposed by Tovo [Tovo 2002] is also discussed. Furthermore, a new method is proposed, i.e. the empirical-$\alpha_{0.75}$ method.

Numerical simulations concerning different spectral densities with simple geometries (constant, linear, parabolic) are used to compare damage estimations from different spectral methods. In addition, they allow us investigating all the correlations existing between rainflow damage and process spectral density.

Chapter 5 discusses bimodal random processes, which have a spectral density formed by the superposition of two well-separated narrow-band contributions. Such spectral

8

densities are typical of loading responses observed in off-shore structures or in automotive chassis components.

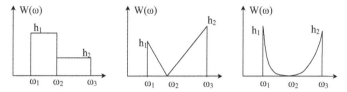

Figure 1.3: Spectral densities used for simulating Gaussian random processes [Chapter 4].

The most used spectral methods specific for bimodal processes are discussed: the single-moment method [Lutes and Larsen 1990, Larsen and Lutes 1991], the Jiao-Moan [Jiao and Moan 1990], the Sakai-Okamura [Sakai and Okamura 1995], the Fu-Cebon [Fu and Cebon 2000] method. Furthermore, an original modification of the Fu-Cebon method (i.e. the modified Fu-Cebon method) is proposed, as an improvement in rainflow damage estimation. Bimodal spectral densities are used in numerical simulations to compare all methods.

In Chapter 6, a simple case study is presented, with the aim to underline all capacities and potentialities of frequency-domain approach in the quick design and sensitivity analysis of real components, especially at early design stages.

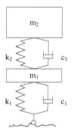

Figure 1.4: Two-degrees-of-freedom system [Chapter 6].

A linear two-degrees-of-freedom model moving on an irregular profile (which is often used to approximate a car-quarter on a road) is considered and the damage associated to the forces acting on the suspension system is investigated, see Figure 1.4.

The linear frequency response analysis is first used to find the response transfer functions of the system, which are then used to find the response spectral density from the input spectral density (which is known from the literature [Dodds and Robson 1973]). Spectral methods provide a quick and easy evaluation of the influence on fatigue damage by the change of one model parameter at a time.

1.2.4. Fatigue analysis of non-Gaussian loadings

A great problem in the frequency-domain analysis of real loads is that the Gaussian assumption is seldom verified. In fact, real systems often exhibit non-linearities, which transform a Gaussian external input into a non-Gaussian internal loading response. In other cases, external loading inputs are non-Gaussian, e.g. wind or wave loads. Deviations from the Gaussian behaviour are detected either by a normal probability plot or by comparing the level-upcrossing spectrum with the Rice's formula. Quantification of the magnitude of the deviation is given by skewness γ_3 and kurtosis γ_4 (a Gaussian loading has $\gamma_3 = \gamma_4 - 3 = 0$).

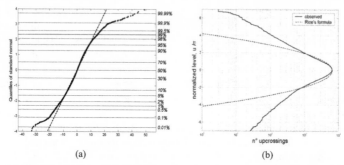

(a)   (b)

Figure 1.5: Non-Gaussian data (force on the bicycle handlebar): (a) normal probability plot (the straight line is the Gaussian distribution); (b) level-upcrossing spectrum [Chapter 8].

The application to non-Gaussian loadings of spectral methods valid for stationary Gaussian processes might underestimate the overall rate of fatigue damage accumulation in non-Gaussian loadings. In addition, such spectral methods are generally calibrated on results given by numerical simulations, which generate stationary loadings, i.e. they do not consider that real loads might possess slight non-stationary characteristics over a quite long time period.

Figure 1.6: Forces measured on the Mountain-bike [Chapter 8].

Consequently, it is quite important to develop spectral methods also valid for non-Gaussian loadings and tested on real time histories.

In Chapter 7, some general properties of the distribution of rainflow cycles are analysed. Then, the TB method developed for Gaussian loadings is further extended to the non-Gaussian case. Subsequently, it is applied to data measured in a Mountain-bike on off-road tracks (see Figure 1.6) and on data from an automotive application.

1.2.5. Comparison with experimental lifetime data

Chapter 9 compares the theoretical predictions with lifetime data obtained in experimental tests, concerning a welded cruciform joint subjected to non-Gaussian random loading [Sarkani et al. 1994, Kihl et al. 1995, Sarkani et al. 1996].

Two types of spectral density are considered, i.e. a broad-band and a bimodal one. Comparison in made amongst rainflow analysis (i.e. rainflow count and linear damage rule) on new simulated time histories, theoretical predictions by the non-Gaussian TB method and experimental lifetime data. Good accuracy in estimations concerning damage from simulations confirms the correctness of the method in including non-normalities, while deviations in respect to experimental data reflect some inadequacy in the linear damage rule.

Chapter 2

COUNTING METHODS FOR VARIABLE AMPLITUDE LOADINGS

2.1. INTRODUCTION

Fatigue is recognised as one of the primary cause of failure of many structures and mechanical components. Since fatigue failures happen without any warning, they can have catastrophic consequences (an example is given in [Whithey 1997]).

The mechanism of fatigue fracture can be explained by the presence of cracks into the metal, which initiate and propagate under the action of the applied load.

A large scatter is usually observed in fatigue strength and it derives from the mechanism of nucleation and propagation of fatigue cracks. In fact, crack initiation is generally localised on the surface of components and it is greatly affected by many different metallurgical factors, as surface defects, microscopic non homogeneities or grain structure. Tests performed with simple loads (e.g. constant amplitude fatigue cycles) show large differences in the fatigue resistance under the same nominal conditions.

In addition, it is also well known that other factors may influence the fatigue resistance, as for example stress concentrations due to specimen geometry, material non homogeneity or defects, or environmental actions (e.g. corrosion).

Further, the complexity also increases with variable amplitude loadings (as irregular loads experienced by real components during service conditions), in which the main problem is the no obvious definition of fatigue cycles. The approach commonly adopted is to first convert the load into a set of fatigue cycles by using a cycle counting method, and then to compute the total damage produced by the load by summing the damage produced by each counted cycle.

The definition of cycle counting methods and suitable damage accumulation rules remain the main issue in the fatigue assessment of irregular loads.

Counting methods are algorithms, which identify fatigue cycles by appropriately combining maxima and minima in the load. Damage accumulation rules are, instead, phenomenological based models for calculating the total damage as an appropriate summation of damage increments associated to each counted cycle.

Amongst all damage rules, the Palmgren-Miner linear damage model is the most popular and used, due to its simplicity. However, experimental investigations using block sequences with different constant amplitude levels (as two-step tests, see [Košút 2002]) or tests with random loads (see [Agerskov and Nielsen 1999, Agerskov 2000]) highlights the presence of interaction and sequence effects, which are not considered by the liner damage accumulation model.

The next sections introduce some basic concepts about the fatigue analysis of variable amplitude loadings. Some counting methods are illustrated and compared.

2.2. FATIGUE ANALYSIS CONCEPTS

A time-varying load applied to a material causes a progressive accumulation of damage that can produce to fracture after a certain amount of time. This macroscopic behaviour is the effect of a complex mechanism on microscopic scale, which is related to migration of dislocations and micro-plastic deformation on some favourable oriented grains.

The accumulation of irreversible plastic deformation, due to dislocation movement, can cause small cracks (micro-cracks) to growth from the surface of the material and to propagate. Small surface defects, as inclusions or geometric defects, can act as points for crack initiation. Several micro-cracks can joint together to form a macroscopic crack (macro-crack), propagating perpendicularly to the applied external load. Failure occurs when the remaining no-cracked area is no longer able to sustain the maximum applied load.

Despite the fatigue phenomenon is generally well understood in its fundamental microscopic aspects, there are no quantitative approaches capable to predict fatigue lifetime and resistance from the microscopic scale. For example, the determination of fatigue strength is usually achieved by experimental tests.

The first systematic investigation of the fatigue phenomenon is due to Wöhler [Schütz 1996]: based on quantitative results obtained on stresses measured on railway axles, he identified stress amplitudes as the most important parameters controlling fatigue life, with tensile mean stresses also having negative effects. In constant amplitude tests, fatigue strength is quantified by the number of cycles to failure, N, under repeated sinusoidal cycles with amplitude s. A different number of cycles to failure is obtained for each amplitude level, with shorter lives associated to larger amplitudes; for many materials, this relation is explicitly given as a straight line in a double-logarithmic diagram (S-N curve):

$$s^k N = C \qquad (2.1)$$

where k and C are material parameters and N is the number of cycles to failure at a given stress amplitude s. For low amplitudes, some materials exhibit a lower limit s_∞ (fatigue limit), below which fracture does not occur, i.e. N tends to infinity; statistical analysis is used in constant amplitude tests in order to estimate k and C, as well as the fatigue limit.

Constant amplitude tests are usually performed with zero-mean cycles; however, the negative effect produced on fatigue life by tensile mean is well known. Beside experimental approaches (i.e. S-N curves at different mean vales), a commonly adopted criterion to account for mean effect is to transform a nonzero-mean cycle into an equivalent damaging cycle, with mean zero and greater amplitude (Goodman correction) [Łagoda et al. 2001]:

$$s_{eq} = \frac{s}{1 - m/S_u} \qquad (2.2)$$

where m is the mean value (see Figure 2.2) and S_u is the ultimate tensile strength.

In the S-N curve there is no information about the process of damage accumulation during cycle repetitions.

In the literature, there are different hypotheses about the way by which damage accumulates during each cycle; a detailed review of different damage accumulation models can be found in [Fatemi and Young 1997].

The simplest hypothesis (Palmgren-Miner damage rule) relates fatigue damage accumulation to the work absorbed by the material during cycle repetitions in a constant amplitude test. Final fracture occurs when total absorbed work reaches a critical level [Miner 1945]; since absorbed work is assumed proportional to the number of cycles, damage accumulates linearly during cycle repetitions and each cycle contributes to the total damage by the fraction:

$$d_i = \frac{1}{N} \qquad (2.3)$$

being N the number of cycles to failure at amplitude s. Damage in time T is calculated as the linear sum of damage contribution from each counted cycle, i.e.:

$$D(T) = \sum_{i=1}^{N(T)} \frac{1}{N_i} \qquad (2.4)$$

where the sum is extended on all $N(T)$ cycles completed in time period T, and fatigue failure occurs when $D(T)$ reaches unity.

The linear rule is load-level and load-sequence independent, i.e. it completely neglects cycle interaction effects often observed in experiments, and which produce crack growth acceleration and/or retardation (for example, the linear rule odes not work in high-low or low-high amplitude sequences [Košút 2002]). The same experimental results given by Miner highlighted that failure could when total damage $D(T)$ is not unity, and it falls in the range between 0.61 and 1.49 (see [Miner 1945]). Similar conclusions are drawn from experimental investigations on welded joints under service loading, see [Agerskov and Nielsen 1999, Agerskov 2000].

2.3. VARIABLE AMPLITUDE LOADINGS: BASIC DEFINITIONS

Some basic properties of an irregular load are of interest, as local extremes (peaks and valleys), ranges (i.e. excursion between consecutive extremes) and level-crossings (see Figure 2.1).

Damage produced by a variable amplitude load is assumed to only depend on the sequence of extremes (maxima and minima), and not on the particular waveform connecting them, i.e. fatigue damage process is assumed rate-independent [Dreßler et al. 1997]. Hence, an equivalent characterisation of a load $x(t)$ is given by the discrete-time sequence of its maxima and minima (sequence of turning points or reversals):

$$TP\big(x(t) \big) = \big\{ x(t_1), x(t_2), x(t_3), x(t_4), x(t_5), x(t_6),... \big\}$$
$$= \big\{ m_0, M_0, m_1, M_1, m_2, M_2,... \big\}$$

where m_k represents a minimum and M_k a maximum.

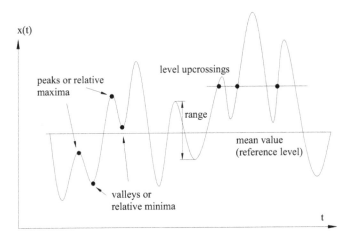

Figure 2.1: Principal parameters of a variable amplitude load.

In practical applications (e.g. digital treatment and analysis of real data), a sequence of turning points can be also discretized into a set of fixed levels (often 64 or 128), leading to a sequence of discretized turning points:

$$DTP\big(x(t) \big) = \big\{ m_0^d, M_0^d, m_1^d, M_1^d, m_2^d, M_2^d,... \big\}$$

Different choices are possible: we can project a turning point to the nearest level, or alternatively to discretize maxima upward and minima downward to the closest discrete level. The first method, giving smaller discretization errors, leads also to smaller error in damage calculation, whereas the second method, giving a discretized load with greater amplitudes than in the original one, leads to a greater (conservative) damage value [Johannesson 1999].

2.4. COUNTING METHODS

A cycle is defined by specifying its highest and lowest points (i.e. its maximum and minimum); a cycle can be thought as a pair $\left(M_k, m_k^*\right) \in \mathfrak{R}^2$, where $M_k \geq m_k^*$, being M_k the maximum (peak) and m_k^* the minimum (valley) associated to M_k. A counting method is a procedure transforming a given load time history $x(t)$ into a set of counted cycles $\left\{\left(M_k, m_k^*\right)\right\}$, where subscript k indicates that the cycle is counted at time t_k. From a fatigue damage point of view, a more conveniently description of a cycle is in terms of its range (or its amplitude) and its mean value:

$$s = \frac{M_k - m_k^*}{2} \qquad \text{amplitude}$$

$$r = M_k - m_k^* = 2s \qquad \text{range} \qquad (2.5)$$

$$m = \frac{M_k + m_k^*}{2} \qquad \text{mean value}$$

which are summarised in Figure 2.2.

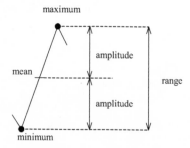

Figure 2.2: Definition of the amplitude, the range and the mean of a cycle.

It is not possible to consider here all counting methods and variants existing in the literature [ASTM 1985, Fuchs et al. 1977]. Hence, we will restrict ourselves to the most used counting methods: peak-valley counting (PVC), level-crossing counting (LCC), range counting (RC), range-pairs counting (RPC), rainflow counting (RFC).

Counting methods differ in the rules by which they pair peaks and valleys in the load to form cycles and half-cycles, even if different rules for the same method may exist (e.g. rainflow count). A possible broad classification of all counting methods is in two classes: one-parameter (or global) methods and two-parameter (or local) methods [Atzori and Tovo 1994].

One-parameter methods (e.g. peak-valley, level-crossing counting) characterise each counted cycle by only one parameter, e.g. its amplitude. Moreover, they also reconstruct cycles by pairing maxima and minima independently of their relative position inside the history, thus disregarding all possible time correlations amongst peaks and valleys defining the set of counted cycles. In other words, one-parameter methods tend to reconstruct cycles not really visible or identifiable in the history, and having a mean value essentially coincident with the reference value of the load. Therefore, the hypothesis that all cycles have the same mean value, coincident with the load reference level, is commonly adopted.

On the contrary, two-parameter methods (e.g. range, range-pairs and rainflow counting) register also the mean value of each counted cycle; in addition, their algorithms also account for the (local) temporal sequence and correlation existing amongst all maxima and minima.

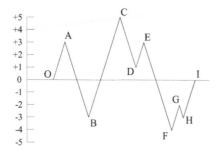

Figure 2.3: Example of a discretized time history.

In the next paragraphs, the counting methods mentioned above are applied to the simple time history shown in Figure 2.3 and are then compared. The reference level of the history, computed as the algebraic mean of all extremes, is zero. All results are collected Table 2.1.

2.4.1. Peak-valley counting (PVC)

The method identifies all peaks above and all valleys below the reference level. The most damaging cycles are counted by taking the highest and the lowest possible peaks and valleys, respectively, until all identified extremes are used. Remaining peaks or valleys can be counted by coupling them with the reference level.

In the example, three peaks (A, C and E) are identified and paired with three valleys (B, F and H); the largest cycle is formed from "+5" (peak C) to "-4" (valley F), and the other two from "+3" to "-3" (A-B and E-H pairs).

The main shortcomings of this method are that not all peaks and valleys in the load are considered (e.g. peaks below and valleys above the reference level); in addition, the method counts all small reversals as much larger cycles.

17

Table 2.1: Results for different counting methods applied to the history of Figure 2.3.

Peak-valley counting		Level-crossing counting	
N° cycles	Range	N° cycles	Range
1	9	1	7.7
1	6	1	6.3
1	6	1	2.8
		1	0.7

Range counting									
N° cycles	0.5	0.5	0.5	0.5	0.5	0.5	0.5	0.5	0.5
Range	8	7	6	4	3	3	2	2	1
Mean	+1	-0.5	0	+3	+1.5	-1.5	+2	-3	-2.5

Range-pairs counting			Rainflow counting			
N° cycles	Range	Mean	N° cycles	Range	Mean	Cycle
1	8	+1	0.5	9	+0.5	$H^+(t_1)$
1	3	+1	0.5	8	+1	$H^-(t_1)$
0.5	3	+0.5	0.5	6	0	$H^+(t_0)$
1	2	1	0.5	4	-2	$H^-(t_4)$
1	1	-2.5	0.5	3	+1.5	$H^-(t_0)$
			1	2	+2	$H(t_2)$
			1	1	-2.5	$H(t_3)$

Range-pairs counting Rainflow counting (case of Repeating Histories)		
N° cycles	Range	Mean
1	9	+0.5
1	6	0
1	2	+2
1	1	-2.5

2.4.2. Level-crossing counting (LCC)

The method identifies cycles based on the level-crossing count of the load, which registers how many times the load crosses a set of fixed levels, counting only positive sloped crossings above and only negative sloped crossings below the reference level (crossings of the reference level are counted when occurring with positive slope).

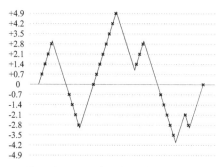

Figure 2.4: Upcrossings for the load in Figure 2.3.

Once the level-crossing count from the load, cycles are formed by identifying the largest possible cycle, followed by the second largest, and so on; all level-crossings are reduced by a unit each time a cycle is formed, until all level crossings are used. In the example, a threshold $\delta = 0.7$ is used, with the largest cycle formed from "+4.9" to "-3.5" crossings, the second cycle from "+2.8" to "-2.8". The remaining crossings form the other two cycles.

2.4.3. Range counting (RC)

This two-parameter method extracts all ranges from a peak to the following valley (max-min count), or from a valley to the next peak (min-max count), counting each range as one half-cycle. The method can also distinguish half-cycles with positive (valley-peak transition) or negative (peak-valley transition) ranges.

2.4.4. Range-pairs counting (RPC)

This method (as described in the ASTM standard [ASTM 1985]) follows the so-called "3-points" algorithm, since three consecutive extremes are considered at a time. These extremes are used to form and to compare two consecutive ranges: X (the range under consideration) and Y (the previous range adjacent to X). A cycle is counted if a range can be paired with a subsequent loading of equal magnitude in the opposite direction. That is, if condition $X \geq Y$ is matched, range Y is counted as a cycle and its extremes are removed from the load, otherwise next peak or valley is read: other two ranges X and Y are then compared, until all ranges in the load are examined.

The remaining uncounted time history (i.e. the residual) is counted backwards by starting from the end; a single range remaining can be counted as half a cycle or one cycle.

The method applied to the example time history is shown in Figure 2.5: ranges OA, DE, GH and BC are counted as cycles and extracted; range FI is the residual. It is worth noting that some portions of the load are not used to form any cycle.

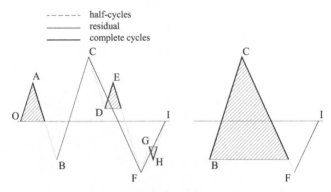

Figure 2.5: Range-pairs counting.

We shall discuss later that method is very similar to the rainflow count, the main difference lying in the treatment of half-cycles, which are handled differently by the two methods.

2.4.5. Rainflow counting (RFC)

This is undoubtedly the most popular and used counting method, since amongst all counting algorithms it has been recognised as the most accurate in identifying damaging events in complex loadings, see [Dowling 1972]. In fact, the method extracts cycles on the basis of the material memory mechanisms, i.e. all cycles correspond to closed hysteresis loops in the material stress-strain plane (see [Dreßler et al. 1997, Anthes 1997]).

In literature one can find different several algorithms defining the rainflow method, based on different rules, which however lead to the same results. In the following sections we shall analyse three algorithms: the "pagoda-roof" method (as originally proposed by Matsuishi and Endo [Matsuishi ad Endo 1968]), the "3-points" algorithm (ASTM rule) [ASTM 1985, Downing and Socie 1982, ESDU] or the alternative "4-points" algorithm (see [Amzallag et al. 1994]), and the non-recursive definition proposed by Rychlik [Rychlik 1987].

In the "pagoda-roof" method, the stress or strain time history is plotted with the time axis vertically downward, with lines connecting peaks and valleys imagined to constitute a series of pagoda roofs; several rules imposed to the rain dripping down these roofs allow cycles and half cycles to be identified and extracted.

The rain flow must start from a peak or valley, in ordered sequence, and it is allowed to drip down, unless:

(a) the rain, starting from a minimum, must stop when it comes opposite to a minimum more negative than minimum from which it initiated; a half cycle is then counted between the initial minimum and the largest peak where it stopped;

(b) the rain, starting from a maximum, must stop when it comes opposite to a maximum more positive than maximum from which it initiated; a half cycle is counted between the initial maximum and the lowest valley where it stopped;

(c) the rain must also stop when it meets the rain from a roof above.

The "pagoda roof" method applied to the example time history is shown in Figure 2.6. Begin at point O and stop opposite valley B (peak A), valley B being more negative than O (half-cycle OA is counted); similarly, rain starting at A must stop opposite peak C (valley B), being C more positive than A (half cycle AB is counted).

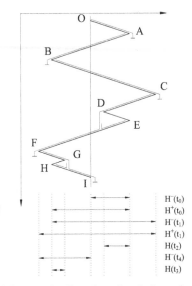

Figure 2.6: The rainflow counting ("pagoda-roof" method), together with Rychlik's method.

Another common rainflow method, described in the ASTM standard, is based on the "3-points" algorithm, since it compares two consecutive ranges at a time (similarly to that described for the range-pairs method.

Range X under consideration and previous range Y adjacent to X are compared; the starting point S is also considered. If the condition $X \geq Y$ is matched, two alternatives are possible: if Y is the first range of the history (i.e. it contains the starting point S), it is counted as a half-cycle and its first extreme is removed, otherwise (i.e. range Y does not contain S), it is counted as a cycle and its extremes (peak and valley) are removed from the history. All ranges that has not been counted are put into the residual and counted as half-cycles.

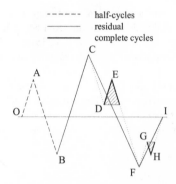

Figure 2.7: Rainflow counting ("3-points"algorithm).

In the example (see Figure 2.7), the first three ranges (ranges OA, AC and BC) are counted as half-cycles (all contain the starting point) and are then removed; ranges DE and GH are counted as cycles; the remaining ranges (residual) are BC, CF and FI.

An alternative (but equivalent) definition of the rainflow count is the "4-points" algorithm [Amzallag et al. 1994], in which four consecutive extremes (e.g. S_1, S_2, S_3 and S_4) are considered at a time, in order to form and to compare three consecutive ranges: $\Delta S_1 = |S_2 - S_1|$, $\Delta S_2 = |S_3 - S_2|$ and $\Delta S_3 = |S_4 - S_3|$. If range ΔS_2 is less then or equal to its adjacent ranges (i.e. $\Delta S_2 \leq \Delta S_1$ and $\Delta S_2 \leq \Delta S_3$), range ΔS_2 is counted as a cycle, its extremes are removed and the remaining parts of the load are connected together. It is immediate to verify in Figure 2.7 that this procedure gives the same result already obtained by previous algorithms.

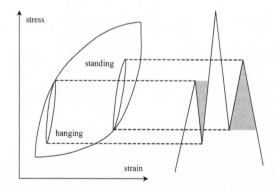

Figure 2.8: Definition of rainflow "standing" and "hanging" cycles.

In the rainflow method, every part of the stress or strain time history is counted once and only once; a half-cycle is always formed from the largest peak to the lowest valley in the history (e.g. range CF in Figure 2.6) and it is included into the residual. Small ranges, which are interruptions of larger loading portions, are counted as full cycles and correspond to closed hysteresis loops in the stress-strain plane (as DE and GH in Figure 2.6).

Rainflow cycles can be further divided into two groups, depending on whether the rainflow minimum occurs before ("standing" cycle) or after ("hanging" cycle) the rainflow maximum (see Figure 2.8).

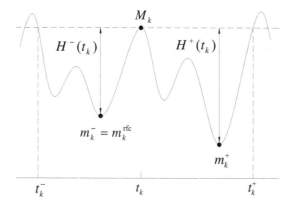

Figure 2.9: Rychlik's definition of the rainflow count (toplevel-up count).

The non-recursive definition of the rainflow count (i.e. toplevel-up count) proposed by Rychlik [Rychlik 1987] gives the same results of previous algorithms, even if it is more tractable from a probabilistic point of view.

Let $x(s)$ be a load observed in time interval $[-T,T]$ and for a maximum $M_k = x(t_k)$ occurring at time t_k (see Figure 2.9) define the following ranges:

$$H^-(t_k) = x(t_k) - \min\{x(s); \; t_k^- < s < t_k \}$$
$$H^+(t_k) = x(t_k) - \min\{x(s); \; t_k < s < t_k^+ \}$$

being t_k^+ the time for the first upcrossing after t of the level $x(t_k)$ (or $t_k^+ = T$ if no such upcrossing exists) and t_k^- the time of the last downcrossing before t_k of the same level (or $t_k^- = -T$ if no such downcrossing exists). Two minima are then defined, i.e. m_k^- and m_k^+ (see Figure 2.9). One cycle or two half-cycles are attached to maximum M_k according to the following rules:

(a) if $H^+(t_k) \geq H^-(t_k)$ and $t_k^- > -T$, or $H^+(t_k) < H^-(t_k)$ and $t_k^+ < T$, a rainflow cycle is counted with range:

23

$$H^{\mathrm{rfc}}(t_k) = \min\big(H^-(t_k), H^+(t_k)\big)$$

(b) if M_k is the first or the last extreme in the load, a half-cycle with range of either $H^+(t_k)$ or $H^-(t_k)$ is counted;

(c) otherwise two half-cycles with ranges $H^+(t_k)$ and $H^-(t_k)$ are counted.

For example, in Figure 2.10 rainflow cycles are attached to maxima occurring at times t_2 and t_3 due to rule (a), a half-cycle is attached to maximum at t_4 (rule (c)), whereas half-cycles are attached to remaining maxima. Figure 2.6 compares results from toplevel-up count with "pagoda-roof" algorithm.

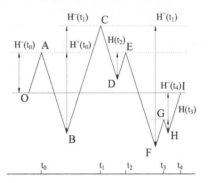

Figure 2.10: Toplevel-up cycle count (Rychlik's method).

2.4.6. Range-pairs and rainflow counts for repeating histories

The rules defining the range-pairs and the rainflow counts seem very similar, but they are not identical. The main differences are in treatment of half-cycles, since all complete cycles are counted similarly by both methods: half-cycles counted by the rainflow method are in fact handled differently by the range-pairs count (compare Figure 2.5 and Figure 2.7).

Rainflow and the range-pair methods give the same results in applications involving repeating loadings, i.e. when a typical segment of a time history is repeatedly applied. In this case, once the largest maximum in the history is reached for the first time, both methods are identical in each subsequent repetition (both methods extract only full cycles and no half-cycles).

According to the ASTM standard [ASTM 1985], the recommendation for the case of repeating loadings is to start counting from the largest peak or the lowest valley in the history; as shown in Figure 2.11, all ranges are counted as full cycles, included the largest one (range CF), and all actually correspond to closed hysteresis loops.

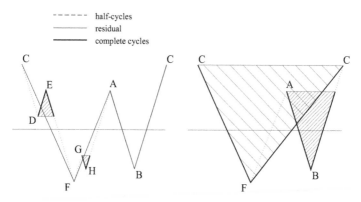

Figure 2.11: Range-pairs and rainflow counts for repeating histories.

Figure 2.12 compares the result of rainflow counting of the example history with the material stress-strain behaviour. As can be seen in Figure 2.12(a), the rainflow counting applied according to the "3-points" algorithm extracts both half-cycles and full cycles: the first three ranges are counted as half-cycles, although from the stress-strain behaviour it would be more reasonable to attach a complete cycle to closed path AB. This apparent discrepancy arises from the fact that path OA starts from the origin and does not actually belong to any stress-strain loop. In this case, the stress-strain behaviour alone is not enough to conclude about range OA behaving like a half-cycle or a complete cycle (for example, the rainflow count treats path OA as a half-cycle). This problem is overcome if the rainflow method is started from the largest peak or the lowest valley (as for the case of repeating histories), where it actually counts all cycles as closed hysteresis loops.

2.5. APPLICABILITY OF COUNTING METHODS TO REAL TIME HISTORIES

All counting methods illustrated in previous sections are different not simply in terms of their algorithms, but also in their capability of correctly identifying fatigue cycles in a real irregular load. It would be of great importance if we were able to identify all possible deficiencies and limits in each method. In fact, some procedures (e.g. peak-valley, range count) have serious drawbacks, which could lead to unreasonable counting results in irregular stress-strain histories observed in real components. According to Dowling [Dowling 1972], only the range-pairs and the rainflow counts seem the best counting methods.

25

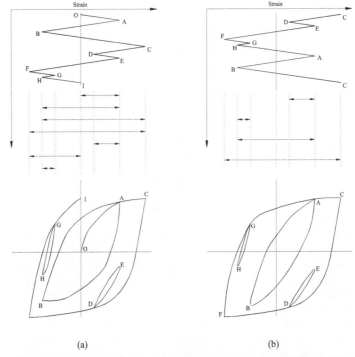

Figure 2.12: Rainflow counting and stress-strain hysteresis loops (a) for the original load of Figure 2.3 and (b) in the case when the load represents a repeating time history.

As an example, Figure 2.13 proposes two sequences: (a) has small reversals superimposed on a large cycle, while (b) derives from (a) by projecting all valleys to the reference level. The peak-valley counting (PVC) predicts the same result for both sequences, even if (b) seems more damaging, and it also counts all small reversals as larger cycles, certainly overestimating total damage. On the contrary, the range count (RC), breaking the largest load variation into small reversals, counts only the smallest ranges and ignores the (quite significant) contribution of the largest cycle, leading to a non-conservative damage estimation. Only the rainflow or the range-pairs, treating all small reversals as interruptions of the largest cycle, extract the largest range in addition to the smallest ones.

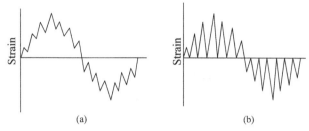

Figure 2.13: Sequences which cause problems for peak-valley and range counting methods.

Counting methods should also account for the mean value effect, when it can potentially affect fatigue life.

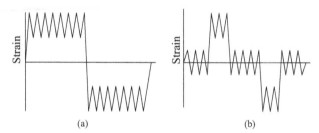

Figure 2.14: Sequences having the same average value taken as the reference level.

Both sequences presented in Figure 2.14 have the same reference level (equal to the algebraic mean of the load). One-parameter methods (e.g. peak-valley and level-crossing counts) predict both sequences to be equally damaging, even if load (b), having more cycles at high tensile levels, seems intuitively more damaging than (a). The alternative of referring to different reference mean levels for contiguous sections seems practicable only for sequences as in Figure 2.14, and would involve quite arbitrarily selection in other cases.

All these shortcomings are avoided if mean value for each counted cycle is determined, as in two-parameter methods (e.g. range, range-pair or rainflow counts).

A further effect often encountered in real loadings is the load interaction effect, which determines variation in fatigue crack growth. If we are using non-linear damage rules, which account for interaction effects in the damage accumulation process, it should be important to account for the correct sequence of counted cycles.

One-parameter methods count cycles ignoring possible time correlations amongst neighbouring extremes, while two-parameter methods (e.g. range-pairs and rainflow counts) are more closely related to the real sequence of extremes. These methods register closed hysteresis loops only at closing time; however, since a large part of a loop can be formed long time before final closing, real damaging effect is not evaluated properly (see Figure 2.11, loop CF).

A modified rainflow count keeping the load sequence is proposed by Anthes [Anthes 1997], where sequence effect has to be considered; the method treats each half-cycle as a damaging event, which may or not form a close loop.

2.6. REPRESENTATION OF A CYCLE COUNT

The result of a cycle count can be represented in many ways, depending on the type and number of parameters evaluated for each counted cycle (that is, one-parameter and two-parameter methods can have different representations).

In engineering practice the simplest representation considers the amplitudes of counted cycles. The result of a cycle count is represented either by an amplitude histogram or a fatigue loading (or cumulative) spectrum $C(s)$, which gives the number of cycles having amplitude higher or equal to s. For a load $x(t)$ in time period $[0,T]$, the loading spectrum is (see Figure 2.15):

$$C(s) = N\left(1 - F(s)\right) \tag{2.6}$$

being N the total number of cycles counted in time interval $[0,T]$, and $F(s)$ the empirical amplitude distribution of counted cycles:

$$F(s) = \frac{\#\left\{t_k : S_k \leq s\right\}}{\#\left\{t_k \in [0,T]\right\}} \qquad , \qquad S_k = \frac{M_k - m_k^*}{2} \tag{2.7}$$

being m_k^* the minimum associated to maximum M_k by a particular counting method (e.g. $m_k^* = m_k^{rc}$ for the range count, or $m_k^* = m_k^{rfc}$ for the rainflow count), and $\#\left\{\cdot\right\}$ is the number of elements in the set $\left\{\cdot\right\}$.

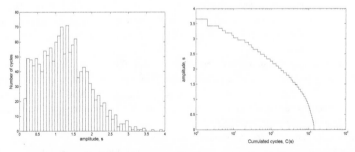

Figure 2.15: Amplitude histogram and loading spectrum for rainflow cycles.

In a given load $x(t)$, each counting method extracts its own set of counted cycles (which may have different amplitudes); hence each method has its own loading spectrum, see for example Figure 2.16.

28

In applications, one often find several definitions of standardised load spectra used to represent typical features of a loading environment for a certain class of structures, as vehicles, car suspension systems, etc.; these spectra are used in experimental load programs for fatigue investigation of service load histories [Schütz et al. 1990, Berger et al. 2002]. The use of standardised loading spectra serves to set up virtual laboratory simulation programs. The idea is to use loading spectra to reproduce realistic tests of real service conditions.

Another alternative one-parameter representation, which does not actually represents the result of a cycle count, is the level-crossing spectrum of a load (see Figure 2.17), which counts how often a level has been reached or exceeded by the load (as an example, Figure 2.4 shows the level upcrossings for the history in Figure 2.3).

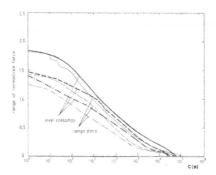

Figure 2.16: Fatigue loading spectra for different counting methods [Schütz et al. 1990].

The level-crossing spectrum gives some important information about possible asymmetries present in the load (for example, it will be used to detect any deviation from the Gaussian hypothesis, see Chapter 8).

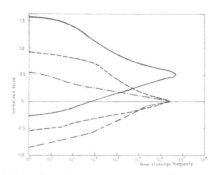

Figure 2.17: Level-crossing spectrum for two loads [Schütz et al. 1990].

When two parameters are considered for each counted cycle (as in two-parameter counts), alternative representations are necessary. Each cycle can be represented either by its peak and valley levels or by its amplitude and mean value. In the first case, cycles are visualised as a cloud of points in the min-max plane (see Figure 2.18 for cycles counted by the range count and rainflow methods).

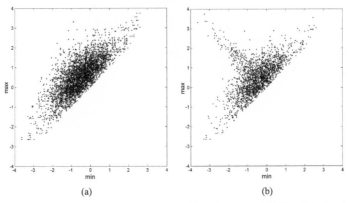

(a)   (b)

Figure 2.18: Comparison of min-max (a) and rainflow (b) counts for a load (4000 cycles): the rainflow method extracts cycles with much larger amplitudes than the range count.

In discretized loads, peak and valley levels (as well as amplitudes and mean values) can take only fixed values, hence the result of a cycle count can be summarised in a matrix. For the rainflow and min-max counts the following matrices are defined:

$$\mathbf{F}^{\text{rfc}} = \left(f_{ij}^{\text{rfc}} \right)_{i,j=1}^{n} \quad , \quad f_{ij}^{\text{rfc}} = \# \left\{ m_r^{\text{rfc}} = u_i , M_r = u_j \right\} \tag{2.8}$$

$$\mathbf{F} = \left(f_{ij} \right)_{i,j=1}^{n} \quad , \quad f_{ij} = \# \left\{ m_r = u_i , M_r = u_j \right\} \tag{2.9}$$

$$\hat{\mathbf{F}} = \left(\hat{f}_{ij} \right)_{i,j=1}^{n} \quad , \quad \hat{f}_{ij} = \# \left\{ M_r = u_i , m_{r+1} = u_j \right\} \tag{2.10}$$

where $\mathbf{F}$ and $\hat{\mathbf{F}}$ are the observed min-max and max-min matrices, and $\mathbf{F}^{\text{rfc}}$ is the observed rainflow matrix. In $\mathbf{F}$ and $\hat{\mathbf{F}}$ matrices, given a maximum M_r, m_r denotes the preceding and m_{r+1} the following minimum. Figure 2.19 and Figure 2.20 show how matrices $\mathbf{F}$ and $\hat{\mathbf{F}}$ are constructed for the short history of Figure 2.3.

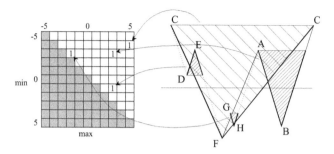

Figure 2.19: Construction of the rainflow matrix $\mathbf{F}^{rfc}$ for a repeating history. Figures are the number of observed full cycles and the grey areas are by definition always zero.

In the min-max count, positive and negative ranges, corresponding to valley-peak and peak-valley transitions, respectively, are collected in two separated matrices, i.e. $\mathbf{F}$ and $\hat{\mathbf{F}}$ matrices. Since $\mathbf{F}$ only has values above the diagonal and $\hat{\mathbf{F}}$ only has values below the diagonal, it is possible to store them as one matrix $\mathbf{F}' = \mathbf{F} + \hat{\mathbf{F}}$, which is often called the "from-to" matrix.

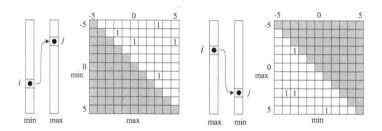

Figure 2.20: The min-max, $\mathbf{F}$, and max-min, $\hat{\mathbf{F}}$, matrices for the time history of Figure 1.3. Figures are the number of observed half-cycles and the grey areas are by definition always zero.

The min-max $\mathbf{F}$ and the max-min $\hat{\mathbf{F}}$ matrices can be converted, after suitable normalisation (i.e. row-sum equal to one), into transition probability matrices, whose elements give the probabilities for min-max and max-min transitions.

For the rainflow count, an asymmetric rainflow matrix can be eventually defined, in which standing cycles are put above and hanging cycles below the main diagonal [Johannesson 1999].

Simple conversion from maximum and minimum to amplitude and mean value can be made according to Eq. (2.5). Figure 2.21 shows a three dimensional representation of a discretized rainflow matrix in terms of amplitude and mean values.

31

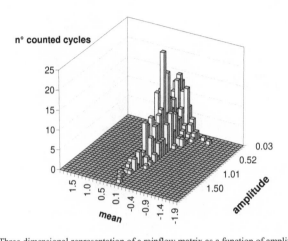

Figure 2.21: Three dimensional representation of a rainflow matrix as a function of amplitude and mean value of counted cycles.

Chapter 3

CYCLE DISTRIBUTION AND FATIGUE DAMAGE IN RANDOM LOADINGS

3.1. INTRODUCTION

Structures and mechanical components during their service life are often subjected to irregular loadings, as those produced by wind, wave or road irregularities.

In order to assess the system service life we should characterise the statistical distribution of the amplitudes and mean values of counted cycles, as well as the total fatigue damage they cause.

Fatigue assessment procedures based on cycle counting schemes and linear damage rules can predict fatigue damage and system service life once the load is known (for example, experimentally). Since all computed quantities are strictly dependent on the measured load, they are random variables. For example both the set of counted cycles and the set of their amplitudes are random variables, as well as the fatigue damage they cause. From measured data we can estimate, for example, unknown parameters of an amplitude distribution with hypothesised shape (e.g. Weibull distribution), e.g. [Wirsching and Sheata 1977, Wu and Huang 1993, Nagode and Fajdiga 1998b]. In practice, the most damaging cycles (not the most frequent) have to be considered, so statistical inference about the damage distribution of counted cycles is preferable [Tovo 2000, Tovo 2001].

In any case, reliable statistical information is achieved only by performing many different measurements, which are often costly and time-consuming. Moreover, in some applications as in the automotive industry, field measurements for a limited period of time serve to estimate the design service life of components; hence, we need to extrapolate measured data towards large cycles' amplitudes [Johannesson and Thomas 2001, Johannesson et al. 2002, Johannesson 2004, Nagode et al. 2001]. Consequently, a single experimental measurement could not be reliable enough to produce satisfactory statistical results.

Other methodologies able to reduce time for data acquisition and analysis, as well as capable to guarantee a complete and reliable statistical description of the random phenomenon under examination, are clearly welcome.

In the so-called spectral methods the irregular loading is modelled as a stationary random process, described by a spectral density in frequency domain. In the most general case, the irregular loading is modelled as a broad-band random process.

The first advantage of spectral methods is the possibility to generate by numerical simulations a large number of time histories having the same spectral density. Another advantage is the possibility to use exact or approximate analytical formulas to relate the statistical distribution of counted cycles and fatigue damage (under a given counting method and a given damage rule) directly to the spectral density of the random process. In this way, a sufficient statistical reliability can be obtained directly in frequency-domain, even from relatively short-term load time-history measurements. On the other hand, the frequency content in a power spectral density summarises all the statistical properties of the random load that are relevant for fatigue.

Since the rainflow count was recognised as the most accurate counting procedure (see Chapter 2), we are mainly interested in describing the statistical distribution of rainflow cycles and in calculating fatigue damage under the linear rule.

A more complete approach would be to find the true expression of the distribution of rainflow cycles and to relate it to the spectral density of the random process, since a simple analytical formula can be used to compute the fatigue damage under the linear rule. However, the complex algorithm which defines the rainflow algorithm makes the relationship between the cycle distribution and the time- (or the frequency-) domain characteristics of the process very complex. Therefore, the true expression of the probability density of rainflow cycles, as well as the correlation between this distribution and the spectral density of the process, are not known at present for the case of broad-band processes.

Some approaches address this problem either with theoretical considerations or by setting completely approximate methods, based on best fitting procedures on many simulation results. Interesting results are also obtained under the Markov hypothesis for the sequence of turning points [Lindgren and Rychlik 1987, Rychlik 1989, Bishop and Sherrat 1990, Frendhal and Rychlik 1993, Bishop 1994]. Simple results are derived (often in approximate form) only for the rainflow damage under the linear rule, with only implicit or no information about the underlying statistical cycle distribution [Wirsching and Light 1980]. In other cases, the amplitude distribution of rainflow cycles is investigated [Dirlik 1985, Zhao and Baker 1992, Tovo 2002].

This Chapter introduces some basic concepts about the characterisation of random processes, both in the time- and in the frequency-domain, as the distribution of peaks and valleys or its spectral density. Some definitions concerning the fatigue analysis of random processes are also introduced, as the statistical characterisation of the distribution of counted cycles and the expected fatigue damage.

3.2. RANDOM PROCESSES AND SPECTRAL DENSITY

Let us model the irregular loading acting in a mechanical component as a stationary and ergodic random process $X(t)$, having a zero mean value. The process is uniquely characterised in time-domain by an autocorrelation function:

$$R_X(\tau) = E\left[X(t)X(t+\tau)\right] \tag{3.1}$$

where $E[\cdot]$ denotes the stochastic mean; alternatively the process is described in frequency-domain by a two-sided spectral density $S_X(\omega)$:

$$S_X(\omega) = \int_{-\infty}^{\infty} R_X(\tau)\, e^{-i\omega\tau}\, d\tau \tag{3.2}$$

We shall often refer to a one-sided spectral density $W_X(\omega)$, defined on positive frequencies only:

$$W_X(\omega) = \begin{cases} 2S_X(\omega) \;, & 0 < \omega < \infty \\[2mm] S_X(0) \;, & \omega = 0 \end{cases} \tag{3.3}$$

A spectral density of a narrow-band process is centred around a restrict range of frequencies, while that of a broad-band process extends over a wider range of frequencies (see Figure 3.1).

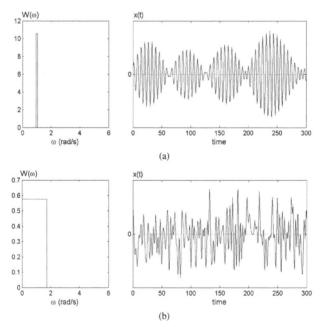

Figure 3.1: Example of (a) narrow-band and (b) broad-band process.

35

Given a one-sided spectral density $W_X(\omega)$, its frequency distribution is uniquely characterised by the set of spectral moments:

$$\lambda_m = \int_0^\infty \omega^m \, W_X(\omega) \, d\omega \qquad m = 1, 2, \ldots \tag{3.4}$$

which represent important time-domain properties of process $X(t)$; for example, the variances for process $X(t)$ and its derivatives $\dot{X}(t)$, $\ddot{X}(t)$ are:

$$\lambda_0 = \sigma_X^2 \ , \qquad \lambda_2 = \sigma_{\dot{X}}^2 \ , \qquad \lambda_4 = \sigma_{\ddot{X}}^2 \tag{3.5}$$

Some other characteristics of process $X(t)$ depend on spectral moments. For example, in a Gaussian process the expected rate of mean upcrossings ν_0 (i.e. mean upcrossings/sec) is:

$$\nu_0 = \frac{1}{2\pi}\sqrt{\frac{\lambda_2}{\lambda_0}} \tag{3.6}$$

and the expected rate of occurrence of peaks ν_p (i.e. peaks/sec) is:

$$\nu_p = \frac{1}{2\pi}\sqrt{\frac{\lambda_4}{\lambda_2}} \tag{3.7}$$

Spectral density $W_X(\omega)$ is also described by bandwidth parameters; the most used are:

$$\alpha_1 = \frac{\lambda_1}{\sqrt{\lambda_0 \lambda_2}} \qquad , \qquad \alpha_2 = \frac{\lambda_2}{\sqrt{\lambda_0 \lambda_4}} \tag{3.8}$$

which belong to a more general family of bandwidth parameters:

$$\alpha_m = \frac{\lambda_m}{\sqrt{\lambda_0 \lambda_{2m}}} \tag{3.9}$$

Note that index m can take also non-integer values; in literature, particular attention has been turned to $\alpha_{0.75}$ bandwidth parameter [Lutes et al. 1984]:

$$\alpha_{0.75} = \frac{\lambda_{0.75}}{\sqrt{\lambda_0 \lambda_{1.5}}} \tag{3.10}$$

The α_m bandwidth parameters are dimensionless numbers, with $0 \le \alpha_m \le 1$, $\alpha_1 \ge \alpha_2$ and $\alpha_{0.75} \ge \alpha_1$ (see Appendix A). Further, $\alpha_m = 0$ if the high-frequency component in the spectral density decays too slowly, such that $\lambda_{2m} = \infty$ (α_2 is then more sensitive to high frequency components than α_1, resulting in its being zero if the spectral density does not decay more rapidly than $|\omega|^{-5}$). An example is the response process $X(t)$ of an oscillator driven by white noise [Lutes and Sarkani 2004, see Example 4.8; Rychlik 1993b], which has $\lambda_4 = \infty$ (i.e. the variance of $\ddot{X}(t)$ is infinite); therefore, $\alpha_1 \approx 1$ (for low damping values) and $\alpha_2 = 0$. Process $X(t)$ also has $v_p = \infty$, i.e. it has infinitely many peaks in any time interval; such a kind of processes are called irregular processes [Rychlik 1993b, Rychlik 1996].

Other spectral parameters for process $X(t)$ may be defined, as spectral parameter q_X [Vanmarcke 1972]:

$$q_X = \sqrt{1 - \frac{\lambda_1^2}{\lambda_0 \lambda_2}} \qquad (3.11)$$

or spectral width parameter ε [Wirsching and Light 1980]:

$$\varepsilon = \sqrt{1 - \frac{\lambda_2^2}{\lambda_0 \lambda_4}} \qquad (3.12)$$

Note that both q_X and ε indexes are related to α_1 and α_2 parameters, i.e. $q_X = \sqrt{1 - \alpha_1^2}$ and $\varepsilon = \sqrt{1 - \alpha_2^2}$; then, it is also $0 \le q_X \le 1$ and $0 \le \varepsilon \le 1$.

In a narrow-band process α_1 and α_2 tend to unity, while for a broad-band process they approach zero. The same is true for the family of α_m parameters, while the opposite obviously holds for q_X and ε.

Some authors suggested to account for other bandwidth parameters relative to the derivative process $\dot{X}(t)$, as for example [Petrucci et al. 2000]:

$$\alpha_{\dot{X}} = \sqrt{\frac{\lambda_4^2}{\lambda_2 \lambda_6}}$$

$$\qquad (3.13)$$

$$q_{\dot{X}} = \sqrt{1 - \frac{\lambda_3^2}{\lambda_2 \lambda_4}}$$

which have the same meaning as α_2 and q_X defined for process $X(t)$. In analogy to Eq. (3.8), we define the bandwidth parameters for the derivative process $\dot{X}(t)$:

$$\beta_1 = \frac{\lambda_3}{\sqrt{\lambda_2 \lambda_4}} \qquad , \qquad \beta_2 = \frac{\lambda_4}{\sqrt{\lambda_2 \lambda_6}} \qquad (3.14)$$

which have analogous meaning as α_1 and α_2 defined for $X(t)$ process. In analogy to Eq. (3.9), the following family of bandwidth parameters for process $\dot{X}(t)$ may be introduced:

$$\beta_m = \frac{\lambda_{m+2}}{\sqrt{\lambda_2 \, \lambda_{2m+2}}} \tag{3.15}$$

It can be proved that $0 \le \beta_m \le 1$ (see Appendix A), then $0 \le \alpha_{\dot{x}} \le 1$ and $0 \le q_{\dot{x}} \le 1$. Note how β_m parameters depend on higher order spectral moments.

An important time-domain characteristic of a process is the irregularity factor, IF, defined as the ratio of the mean upcrossing, ν_0, to the peak, ν_p, intensity:

$$IF = \frac{\nu_0}{\nu_p} \tag{3.16}$$

As α_1 and α_2 bandwidth parameters, also IF ranges from zero (broad-band processes) to unity (narrow-band processes). For Gaussian processes, explicit expressions for both ν_0 and ν_p show that IF equals α_2, which so gains another possible interpretation [Lutes and Sarkani 2004]. In a non-Gaussian process, we can not expect IF and α_2 to be identical, but they will generally be quite similar.

We conclude this section with other properties of process $X(t)$, which are important in fatigue analysis, i.e. the level-crossing spectrum and the peak distribution.

If process $X(t)$ is Gaussian, its level-crossing spectrum is given by the well-known Rice's formula [Lutes and Sarkani 2004], i.e.:

$$\nu(u) = \nu_0 \, e^{-\frac{u^2}{2\lambda_0}} \tag{3.17}$$

where ν_0 (i.e. the maximum of the level-crossing spectrum crossing intensity) is the expected mean upcrossing rate, Eq. (3.6).

If process $X(t)$ is a zero-mean Gaussian process, the probability density of peaks is [Lutes and Sarkani 2004]:

$$p_p(u) = \frac{\sqrt{1-\alpha_2^2}}{\sqrt{2\pi}\,\sigma_X} e^{-\frac{u^2}{2\sigma_X^2(1-\alpha_2^2)}} + \frac{\alpha_2 u}{\sigma_X^2} e^{-\frac{u^2}{2\sigma_X^2}} \Phi\!\left(\frac{\alpha_2 u}{\sigma_X \sqrt{1-\alpha_2^2}}\right) \tag{3.18}$$

while its cumulative distribution function is:

$$P_p(u) = \Phi\!\left(\frac{u}{\sigma_X \sqrt{1-\alpha_2^2}}\right) - \alpha_2 \, e^{-\frac{u^2}{2\sigma_X^2}} \Phi\!\left(\frac{\alpha_2 u}{\sigma_X \sqrt{1-\alpha_2^2}}\right) \tag{3.19}$$

being $\Phi(\cdot)$ the standard normal distribution function:

$$\Phi(u) = \frac{1}{\sqrt{2\pi}} \int_{-\infty}^{u} e^{-\frac{t^2}{2}} \, dt \qquad (3.20)$$

Figure 3.2 shows density $p_p(x)$ for different values of the α_2 parameter.

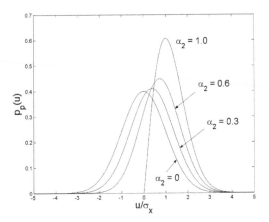

Figure 3.2: Peak distribution for different values of α_2 parameter.

The density function of valleys is symmetrical with that of peaks, $p_v(v) = p_p(-v)$, and its cumulative distribution is:

$$P_v(v) = \Phi\left(\frac{v}{\sigma_X \sqrt{1-\alpha_2^2}}\right) + \alpha_2 e^{-\frac{v^2}{2\sigma_X^2}} \left[1 - \Phi\left(\frac{\alpha_2 v}{\sigma_X \sqrt{1-\alpha_2^2}}\right)\right] \qquad (3.21)$$

For strictly narrow-band processes ($\alpha_1 = \alpha_2 = 1$), the peak distribution turns into a Rayleigh distribution (defined for $x \geq 0$), having the following probability and cumulative distributions:

$$r(x) = \frac{x}{\sigma_X^2} e^{-\frac{x^2}{2\sigma_X^2}} \quad , \quad R(x) = 1 - e^{-\frac{x^2}{2\sigma_X^2}} \qquad (3.22)$$

For processes having mean value different from zero, equation above can be easily updated by means of a variable shift (see [Lutes and Sarkani 2004]).

Based on Figure 3.1 and Figure 3.2, we observe that a strictly Gaussian narrow-band process has all peaks above zero (i.e. no negative peaks); on the opposite, a broad-band

39

Gaussian process has both positive and negative peaks. In fact, from Eq. (3.19) we can see that the fraction of peaks below zero is:

$$P_p(0) = \frac{1-\alpha_2}{2} \tag{3.23}$$

and this fraction is obviously zero for in a strictly narrow-band process ($\alpha_2 = 1$).

3.3. PROPERTIES OF THE DISTRIBUTION OF COUNTED CYCLES

Fatigue damage is related to amplitudes and mean values of loading cycles. In a variable amplitude load, a counting method extracts cycles by pairing peaks and valleys. The set of counted cycles depends on the load examined, thus, if the load is modelled as a random process $X(t)$, it is a set of random variables.

The fundamental problem in the fatigue assessment framework is to find, for a random process $X(t)$, the true distribution of counted cycles under a chosen counting method, e.g. the rainflow count, since fatigue damage under the linear damage rule actually depends on this cycle distribution. In addition, the aim is also to establish the proper correlation between the distribution of rainflow cycles and the spectral density of process $X(t)$.

Therefore, the distribution of rainflow cycles plays a fundamental role in the fatigue analysis of random process $X(t)$. However, both the true expression of this distribution and its explicit correlation with the process spectral density are not known at present.

In next sections, the statistical distribution of counted cycles is characterised by two alternative (but related) descriptors, i.e. the count intensity $\mu(u,v)$ (defined by Rychlik [Rychlik 1993c]) and the joint probability density function $h(u,v)$.

Let $x(t)$, $0 \leq t \leq T$, a time history taken from random process $X(t)$. Let us suppose that a counting method gives a finite set of $N(T)$ counted cycles $\{(M_k, m_k^*)\}$, where M_k and m_k^* are the maximum and the minimum of each cycle, with $m_k^* < M_k$, and where can be for example $m_k^* = m_k^{rfc}$ for the rainflow count or $m_k^* = m_k^{rc}$ for the range (max-min) count. Let $N_T(u,v)$ be the number of cycles counted in $x(t)$ such that the maximum M_k is higher than u and the attached minimum m_k^* is lower than v, i.e.:

$$N_T(u,v) = \#\left\{ (M_k, m_k^*) : M_k > u \geq v > m_k^* \right\} \tag{3.24}$$

where $\#\{\cdot\}$ is the number of elements in the set $\{\cdot\}$. Function $N_T(u,v)$ is called the count distribution. Further, let $\mu_T(u,v)$ denote the expected value of $N_T(u,v)$, and define the count intensity as:

$$\mu(u,v) = \lim_{T \to \infty} \frac{\mu_T(u,v)}{T} \quad \text{with} \quad \mu_T(u,v) = E\left[N_T(u,v) \right] \tag{3.25}$$

40

assuming the limit exists (for ergodic processes the limit exists). Note that for stationary loads:

$$\mu_T(u,v) = T\,\mu(u,v) \qquad (3.26)$$

The expected count $\mu_T(u,v)$ can be thought as the asymptotic shape of the counting distribution $N_T(u,v)$ for process $X(t)$.

Note that $\mu_T(u,v)$ is an increasing function of v and a decreasing function of u, i.e. for any fixed $u \geq v$:

$$\mu_T(u,v) \leq \min\big(\mu_T(u,u), \mu_T(v,v)\big) = \mu_T^+(u,v) \qquad (3.27)$$

Thus, the count intensity is bounded by its diagonal values. This property will be subsequently used to construct an upper bound for fatigue damage.

According to the above definition $\mu_T(u,u)$ is the number of cycles (M_k, m_k^*) such that $M_k > u > m_k^*$. Let us define a cycle count as a crossing-consistent method if $\mu_T(u,u)$ equals the expected number of u-upcrossings of process $X(t)$, i.e.:

$$\mu_T(u,u) = \#\big\{\ t_k :\ t_k \text{ is a } u\text{-upcrossing of } x(t)\ \big\} \qquad (3.28)$$

For simplicity of notation, we write $\mu_T(u,u) = \mu_T(u)$. In a stationary process, $\mu_T(u) = T\,v(u)$, where $v(u)$ is the level-crossing spectrum. If the process is also Gaussian, the level-crossing spectrum is given by Eq. (3.17) (Rice's formula), thus the expression of the number of u-upcrossings in time period T is:

$$\mu_T(u) = T\,\frac{1}{2\pi}\sqrt{\frac{\lambda_2}{\lambda_0}}\,e^{-\frac{u^2}{2\lambda_0}} \qquad (3.29)$$

Let us also define a counting method as a complete counting if it pairs every peak in a load with a lower or equal valley, so that the number of all counted cycles is equal to the total number of peaks. Note that a complete counting method is also crossing-consistent [Tovo 2002].

Amongst usual counting methods, the range (min-max) count is both complete and crossing-consistent, the level-crossing is clearly crossing-consistent, whereas the peak-valley count is neither complete nor crossing-consistent. The rainflow count needs instead some cautions, since it can be defined according to different algorithms, and residuals (i.e. unclosed loops) often occur in several definitions. However, modifications can be adopted to count only full cycles (e.g. case of repeating histories); thus, the rainflow count is actually a complete method.

We introduce now an alternative (and complementary) descriptor of the statistical distribution of counted cycles based on probability density concept.

41

Let $h(u,v)$ be the joint distribution of cycles counted in random process $X(t)$, as a function of peak u and valley v levels; note that $h(u,v)$ is null for $u < v$. The related cumulative distribution function:

$$H(u,v) = \int_{-\infty}^{u} \int_{-\infty}^{v} h(x,y) \, dx \, dy \qquad (3.30)$$

gives the probability to count a cycle with peak lower or equal to level u and valley lower or equal to level v. By a simple change of variables, we can express the cycle distribution in terms of amplitudes and mean values:

$$p_{a,m}(s,m) = 2 \, h(m+s, m-s) \qquad (3.31)$$

while the distribution of counted cycles as a function of amplitudes is given by the marginal probability density function:

$$p_a(s) = \int_{-\infty}^{+\infty} p_{a,m}(s,m) \, dm \qquad (3.32)$$

In a complete counting method, a cycle is attached to each maximum in the history, thus the expected intensity of counted cycles, say ν_a, is clearly equal to the expected intensity of peaks, i.e. $\nu_a = \nu_p$. Similarly, since due to completeness every peak is paired with an equal or lower valley, the marginal distribution of cycles must be also related to the distributions of peaks and valleys [Tovo 2002]:

$$\begin{cases} p_p(u) = \int_{-\infty}^{u} h(u,v) \, dv \\[2em] p_v(v) = \int_{v}^{+\infty} h(u,v) \, du \end{cases} \qquad (3.33)$$

This relation is very general and holds for both Gaussian and non-Gaussian processes, when proper peak and valley distribution are considered.

Every counting method defines its own probability density of counted cycles, $h(u,v)$, but, if the method is complete, its distribution must satisfy Eq. (3.33). Therefore, also the distribution associated to the rainflow count (which is a complete procedure) is a solution of the above equations.

Since the above equations are linear any linear combination of two solutions is also a solution. This condition will be used in subsequent sections to construct a method for estimating the distribution of rainflow cycles.

For a given counting method, both $H(u,v)$ and $\mu(u,v)$ functions are two alternative cumulative distributions, which are related as:

$$\mu(u,v) = v_a \int_{x=u}^{+\infty} \int_{-\infty}^{y=v} h(x,y) \, dx \, dy \qquad (3.34)$$

where in complete counting methods we can put $v_a = v_p$.

Another important descriptor commonly used in the engineering field is the cumulative (or loading) cycle spectrum, defined as the percentile number of cycles having amplitudes higher than or equal to s, i.e.:

$$F(s) = \int_s^{+\infty} p_a(x) \, dx \qquad (3.35)$$

where $p_a(s)$ is the probability density of the amplitude s of counted cycles.

3.4. FATIGUE DAMAGE

Once constant amplitude fatigue properties are given (as a S-N curve), the knowledge of the statistical distribution of cycles counted in random process $X(t)$ allows calculating the total fatigue damage under the Palmgren-Miner rule. In the next, we will compute fatigue damage based on both the marginal amplitude distribution $p_a(s)$ and the expected count $\mu_T(u,v)$.

In time history $x(t)$, $0 \le t \le T$, the total damage under the linear damage rule is:

$$D(T) = \sum_{i=1}^{N(T)} \Delta D_i = \sum_{i=1}^{N(T)} \frac{1}{N_i} \qquad (3.36)$$

being ΔD_i the damage increment associated to each i-th counted cycle (see Chapter 2), N_i the number of cycles to failure associated to stress amplitude s_i, and $N(T)$ the number of all counted cycles. Since damage increment ΔD_i depends on constant amplitude fatigue properties by the S-N curve, previous formula becomes:

$$D(T) = \sum_{i=1}^{N(T)} \frac{s_i^k}{C} \qquad (3.37)$$

where s_i is the amplitude of the i-th counted cycle and $s^k N = C$ is the S-N curve. Note that the S-N curve used in Eq. (3.37) is a simplified model which does not account for both the fatigue limit s_∞ and the static strength S_u.

In a random process, s_i is obviously a random variable and its distribution clearly depends on the process itself and on the counting method used (e.g. rainflow count), thus total damage $D(T)$ is a random variable too. More precisely, damage $D(t)$ as a function of time variable t should be modelled as a non-decreasing random process (it should be more correctly defined as a rate-independent functional defined on $X(t)$).

It is very difficult to find the exact distribution of $D(T)$, also in the validity of linear damage accumulation hypothesis. In fact, even if the probabilistic structure of process $X(t)$ is well defined, damage is a complicate non-linear functional defined on $X(t)$. Consequently, we will mainly concentrate on the expected value of the fatigue damage and the system service life (where damage is evaluated under the linear damage accumulation rule). It is worth noting that some approximations adopt a normal distribution for damage [Kececioglu et al. 1998], often under the Markov assumption for the sequence of extremes [Rychlik et al. 1995]. Assuming a non-linear damage rule, more complex theories can be developed [Rejman and Rychlik 1993, Košút 2004].

If the number $N(T)$ of counted cycles is large, and assuming the hypothesis that amplitudes have the same distribution and that the dependence between them is weak (i.e. all amplitudes are assumed to be independent and identically distributed), the expected fatigue damage for process $X(t)$ is calculated by taking expectation of Eq. (3.37) [Madsen et al. 1986, Mood et. al. 1987]:

$$\overline{D}(T) = E[D(T)] = E\left(\sum_{i=1}^{N(T)} \frac{s_i^k}{C}\right) = E[N(T)]\frac{E[s^k]}{C} \qquad (3.38)$$

writing $\overline{N} = E[N(T)]$ for the expected number of cycles counted in time T. This result is valid independently of the counting method considered.

In stationary processes, $\overline{N} = v_a T$, where v_a is the expected intensity of counted cycles; in complete counts (as the rainflow method), $v_a = v_p$, where v_p uniquely depends on the spectral density of process $X(t)$, see Eq. (3.7). Furthermore, in stationary processes we have that $\overline{D}(T) = T\,\overline{D}(1)$, where $\overline{D}(1)$ is called the expected damage intensity (i.e. the damage per time unit).

The formula in Eq. (3.38) shows that the expected damage $\overline{D}(T)$ (or, equivalently, the damage intensity $\overline{D}(1)$), is related to k-th moment of the amplitude distribution, $p_a(s)$. By neglecting mean value effect, to simplify matter, the explicit formula for calculating the expected damage intensity is:

$$\overline{D}(1) = v_a\,C^{-1}\int_0^{+\infty} s^k\,p_a(s)\,ds \qquad (3.39)$$

In the following, we shall write $\overline{D}(1) = \overline{D}$ for the damage intensity.

Formula above clearly states that, for a given random process $X(t)$ (i.e. for a given spectral density), damage intensity uniquely depends on the counting method adopted though the distribution $p_a(s)$, or equivalently $h(u,v)$. We shall refer in particular to the rainflow counting method, since amongst all algorithms it is recognised as the best one (see Chapter 2). Therefore, distribution $h_{RFC}(u,v)$ of rainflow cycles gives the rainflow damage intensity under the linear rule, $\overline{D}_{RFC}$, through its marginal distribution $p_a^{RFC}(s)$ substituted into Eq. (3.39).

The above discussion highlights that rainflow distribution $h_{RFC}(u,v)$ plays a fundamental role in the rainflow fatigue analysis of process $X(t)$. Unfortunately, because of the complicate algorithm which defines the rainflow method, at present no explicit closed-form analytical solution for $h_{RFC}(u,v)$ density, as well as for rainflow damage $\overline{D}_{RFC}$, is available when $X(t)$ is a broad-band process.

Damage $D(T)$ may be alternatively related to the cycle distribution properties of process $X(t)$ by the use of the counting distribution $N_T(u,v)$, as suggested by Rychlik [Rychlik 1993b].

Let us consider a cycle (u,v) having maximum and minimum at levels u and v, respectively, and denote by $d(u,v)$ the damage it causes, according to the S-N curve, i.e. $d(u-v) = C^{-1}(u-v)^k$. In the hypothesis that $d(0) = 0$ and that $N_T(u,u)$ is a bounded function of u, an integration by parts shows that total damage $D(T)$, under the linear damage rule, is finite and given by [Frendhal and Rychlik 1993]:

$$D(T) = d'(0) \int_{-\infty}^{+\infty} N_T(u,u)\,du + \int_{-\infty}^{+\infty} \int_{-\infty}^{u} N_T(u,v)\,d''(u-v)\,dv\,du \qquad (3.40)$$

This integral damage formulation links the total damage to the cumulative count distribution $N_T(u,v)$; by replacing the counting distribution $N_T(u,v)$ by its expectation $\mu_T(u,v)$, we obtain the formula for the expected damage $\overline{D}(T)$ [Rychlik 1993b], i.e.:

$$\overline{D}(T) = d'(0) \int_{-\infty}^{+\infty} \mu_T(u,u)\,du + \int_{-\infty}^{+\infty} \int_{-\infty}^{u} \mu_T(u,v)\,d''(u-v)\,dv\,du \qquad (3.41)$$

In other words, while the counting distribution $N_T(u,v)$ defines the total damage, $D(T)$, the expected count $\mu_T(u,v)$ defines the expected damage, $\overline{D}(T)$, and correspondingly the counting intensity $\mu(u,v)$ gives the expected damage intensity, $\overline{D}$.

Furthermore, if $d''(s) \geq 0$ for all amplitudes $s = u - v \geq 0$, one can use Eq. (3.41) to compare damages from different counting procedures, based on their distribution properties.

Let us consider in particular the range and the rainflow counts. On the basis of their algorithms, given maximum M_k of a generic counted cycle, the attached rainflow counted minimum m_k^{rfc} is always equal or lower than the attached minimum counted by the range count, m_k^{rc}. in particular, for a given maximum M_k, it is always $m_k^{rfc} \leq m_k^{rc}$, m_k^{rfc} being the rainflow minimum, and m_k^{rc} the range count minimum, from which it follows that $\mu_T^{rc}(u,v) \leq \mu_T^{rfc}(u,v)$. In addition, as stated by Eq. (3.27), the counting distribution for a crossing-consistent method is always bounded by its diagonal values, thus we can write:

$$\mu_T^{rc}(u,v) \leq \mu_T^{rfc}(u,v) \leq \mu_T^{+}(u,v) \qquad (3.42)$$

Based on Eqs. (3.41) and (3.42), we conclude that damage using rainflow count always bounds the damage obtained using the range count. Furthermore, an upper bound, say $D^+(T)$, for all crossing-consistent counting methods exists, i.e.:

$$D_{RC}(T) \leq D_{RFC}(T) \leq D^+(T) \tag{3.43}$$

Equation above is valid for a load defined in time interval $[0,T]$ and it equivalently holds for damage intensities as well. We will see later on that the upper bound $D^+(T)$ coincides for Gaussian loads with the narrow-band approximation.

Concluding this section, we propose some considerations about fatigue failure time. Since fatigue damage $D(T)$ is a random variable, fatigue failure time T_f is a random variable too.

As for the fatigue damage, the problem of the probabilistic characterisation of the fatigue failure time is very complex and then we mainly focus on expected values.

Fracture is assumed to occur when damage $D(T)$ reaches a critical level D_{cr} (often $D_{cr} = 1$). Since fatigue failure time T_f is defined by $D(T_f) = D_{cr}$, the failure probability is given by:

$$P\left(T_f \leq T\right) = P\left(D(T) \geq D_{cr}\right) \tag{3.44}$$

Limiting ourselves to expected values, if $D(T)$ has a relatively small variance, then it is concentrated around its mean value and the simplest estimator of T_f, say $\hat{T}_f$, is the solution of the moment equation:

$$\overline{D}(\hat{T}_f) = D_{cr} \tag{3.45}$$

In stationary loadings $\overline{D}(\hat{T}_f) = \hat{T}_f \cdot \overline{D}(1)$, then $\hat{T}_f$ is given by:

$$\hat{T}_f = \frac{D_{cr}}{\overline{D}(1)} \tag{3.46}$$

where $\overline{D}(1)$, the damage intensity, is given by Eq. (3.39).

Chapter 4

FATIGUE ANALYSIS OF GAUSSIAN RANDOM LOADINGS

4.1. INTRODUCTION

Chapter 2 introduced some basic concepts related to the fatigue analysis of variable amplitude loadings (e.g. deterministic loadings). Schemes based on counting methods and damage accumulation rules are generally used for the fatigue analysis and service life prediction of structure and mechanical components subjected to irregular loadings.

The previous Chapter extended the concept valid for the case of deterministic loads to the case of random loadings and it discussed some general properties of stationary random processes (e.g. the concept of spectral density), as well as the probabilistic description of the distribution of rainflow cycles and their fatigue damage under the linear rule. In particular, some fundamental quantities, as the joint distribution $h_{RFC}(u, v)$ or the marginal amplitude distribution $p_{a,m}(s)$ of rainflow cycles and the expected rainflow fatigue damage $\overline{D}_{RFC}$, were introduced.

The hypothesis commonly adopted in practice is that random processes are Gaussian. In this Chapter we consider in more detail the most commonly used spectral methods for fatigue analysis of broad-band Gaussian random processes: the peak-approximation (PV), the narrow-band approximation (NB) and the Wirsching-Light (WL) correction formula, the Dirlik (DK) approximate amplitude density, the Zhao-Baker (ZB) model, the Petrucci-Zuccarello (PZ) approximate damage. Then, other two original methods will be presented, i.e. the Tovo-Benasciutti (TB) method and the empirical-$\alpha_{0.75}$ method.

Methods mentioned above differ in the quantities that are estimated: some of them only estimate the rainflow damage $\overline{D}_{RFC}$ (with no interest on the underlying cycle distribution), while others also assess the marginal amplitude density $p_{RFC}(s)$, the fatigue damage being computed as in Eq. (3.39). Only the TB method also estimate the joint distribution $h_{RFC}(u, v)$, which gives the possibility to extend the applicability also to non-Gaussian loadings. The fatigue analysis of non-Gaussian random processes will be addressed in Chapter 7.

47

Table 4.1: Analogies and differences amongst all spectral methods considered in this Chapter.

Spectral method	Spectral parameters required		Information supplied		
	$\lambda_0, \lambda_2, \lambda_4$	Other parameters	Joint density $h_{RFC}(u,v)$ or $p_{RFC}(s,m)$	Marginal amplitude density $p_{RFC}(s)$	Damage
NB	$\times$			$\times$	$\times$
WL	$\times$				$\times$
DK	$\times$			$\times$	$\times$
ZB (simplified)	$\times$			$\times$	$\times$
ZB (improved)	$\times$	$\times$		$\times$	$\times$
PZ	$\times$	$\times$			$\times$
TB	$\times$	$\times$	$\times$		$\times$
Empirical-$\alpha_{0.75}$		$\times$			$\times$

4.2. ANALYTICAL SOLUTIONS FOR FATIGUE DAMAGE ASSESSMENT

The main problem in the fatigue damage assessment procedure is the estimation of the expected rainflow damage intensity $\overline{D}_{\mathrm{RFC}}$ for random process $X(t)$ and its related fatigue life. As pointed out in the previous Chapter, this problem can be solved by first estimating the true rainflow cycle distribution $h_{\mathrm{RFC}}(u,v)$ and then computing damage intensity $\overline{D}_{\mathrm{RFC}}$ under the linear damage hypothesis; alternatively, direct estimation of fatigue damage is possible as well.

Methods addressing this problem can be divided essentially into few categories. Some of them first estimate the true rainflow cycle distribution (as the joint density, $h_{\mathrm{RFC}}(u,v)$, or its marginal density, $p_{\mathrm{a}}(s)$) and then compute damage under the linear rule according to Eq. (3.39) [Dirlik 1985, Zhao and Baker 1992, Tovo 2002]. Other methods, instead, give exact or approximate formulas for directly estimating rainflow damage $\overline{D}_{\mathrm{RFC}}$, without information about the underlying cycle distribution [Wirsching and Light 1980, Petrucci and Zuccarello 2001]. Finally, other methods estimate the rainflow damage by adopting the Markov hypothesis for the sequence of extremes [Frendhal and Rychlik 1993]. Table 4.1 gives an overall comparison among all methods.

A further goal is also to establish the dependence existing between the rainflow cycle distribution (or rainflow damage) and the frequency-domain characteristics of process $X(t)$, synthesised by its power spectral density, and specifically to investigate the true set of bandwidth parameters involved in this dependence.

In the next we shall give a brief review of the methods applicable to stationary Gaussian random processes; a complete survey can be found in [Bouyssy et al. 1993]. Other approaches specifically developed for Gaussian processes with a bimodal spectral density will be illustrated in Chapter 5, while the fatigue analysis of non-Gaussian loadings is considered in Chapter 7.

4.2.1. Peak approximation

The fatigue damage in process $X(t)$ can be estimated under the peak-valley counting assumption, in which each peak level determines the corresponding cycle amplitude; the amplitude distribution is then estimated according to the peak distribution and damage intensity becomes [Tovo 2002]:

$$\overline{D}_{\mathrm{PV}} = \nu_{\mathrm{p}}\, C^{-1} \int\limits_{0}^{+\infty} s^{k}\, p_{\mathrm{p}}(s)\, \mathrm{d}s \qquad (4.1)$$

For Gaussian processes, the peak distribution $p_{\mathrm{p}}(s)$ is given by Eq. (3.18) and consequently the damage intensity is given by two contributions:

$$\overline{D}_{PV} = v_p\, C^{-1} \left[\ \int\limits_{0}^{+\infty} s^k\ \frac{\sqrt{1-\alpha_2^2}}{\sqrt{2\,\pi}\,\sigma_X}\ e^{-\frac{s^2}{2\sigma_X^2(1-\alpha_2^2)}}\ ds\ + \right.$$

$$\left. +\ \int\limits_{0}^{+\infty} s^k\ \frac{\alpha_2\, s}{\sigma_X^2}\ e^{-\frac{s^2}{2\sigma_X^2}}\ \Phi\!\left(\frac{\alpha_2\, s}{\sigma_X\,\sqrt{1-\alpha_2^2}}\right) ds\ \right] \tag{4.2}$$

Integration limits clearly consider only contribution from positive peaks. This approach has been called the peak approximation of fatigue damage [Lutes and Sarkani 2004].

It is worth noting that the peak-valley counting is not a complete procedure; in fact in calculating fatigue damage in broad-band loadings it completely neglects the fraction of negative peaks. In addition, as in the deterministic case, the peak-valley count generally overestimates total damage (particularly for irregular loadings), its damage being greater even than damage of narrow-band approximation, as shown in [Tovo 2002].

On the basis of these considerations, although Lutes and Sarkani support the peak approximation [Lutes and Sarkani 2004], in our opinion is not useful to estimate fatigue damage through the integral deriving from the peak-count as reported in Eq. (4.2), as done in [Lu and Liu 1997, Lu et al. 1998] and with an improved version in [Lu and Jiao 2000]. In fact, this approach has been shown to give a damage predictor always above the upper damage value estimated by any complete and crossing consistent counting method (see [Tovo 2002]).

We note that another simple approach approximating the rainflow amplitude distribution through the peak distribution is proposed in the literature [Kim & Kim 1994], even if it gives not satisfactory results (see [Petrucci and Zuccarello 1999]).

4.2.2. Narrow-band approximation

For a strictly narrow-band Gaussian process $X(t)$ (see Figure 4.1) it is reasonable to assume the amplitude distribution $p_a(s)$ coincident with the peak distribution $p_p(x)$, which in a narrow-band process is Rayleigh, see Eq. (3.22). Furthermore, the intensity of counted cycles, v_a, can be taken equal to the mean upcrossing intensity, v_0, given by Eq. (3.6). Then, by calculating the fatigue damage intensity as in Eq. (3.39) one finds:

$$\overline{D}_{NB} = v_0\, C^{-1} \left(\sqrt{2\,\lambda_0}\right)^k \Gamma\!\left(1 + \frac{k}{2}\right) \tag{4.3}$$

where $\Gamma(\cdot)$ is the Gamma function. Equation above is valid for a S-N curve with a single slope over the whole range of amplitudes; a closed-form solution including a slope change is developed in [Tunna 1986]. An extension of the narrow-band approximation to the case of a non-linear damage rule can be found in [Wu et al. 1997, Liou et al. 1999].

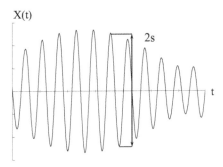

Figure 4.1: Sample of a narrow-band process.

The previous expression holds exactly only for a strictly narrow-band process $X(t)$. When instead it is applied to a process $X(t)$ that is broad-band, the predicted damage value is that of an equivalent ideal narrow-process, with the same variance and a number of peaks equal to the number of upcrossings (or downcrossings) of the mean level of real broad-band process $X(t)$. This is the so-called narrow-band (or Rayleigh) approximation of the fatigue damage of a broad-band process and has frequently been used in engineering applications [Madhavan Pillai and Meher Prasad 2000, Gobbi and Mastinu 1998, Gobbi and Mastinu 2000].

It is widely accepted the fact that the narrow-band approximation, when applied to wide-band processes, tends to overestimate the rainflow fatigue damage; this was proved rigorously by Rychlik [Rychlik 1993a].

Hence, some authors proposed to approximate the rainflow damage intensity $\overline{D}_{\mathrm{RFC}}$ by correcting (namely, by reducing) the damage value predicted by the narrow-band approximation [Wirsching and Light 1980]:

$$\overline{D}_{\mathrm{RFC}} = \rho_{\mathrm{WL}} \, \overline{D}_{\mathrm{NB}} \tag{4.4}$$

in which ρ_{WL} is an empirical correction factor assumed to be dependent on the fatigue curve parameters and on α_2 bandwidth parameter:

$$\rho_{\mathrm{WL}} = a(k) + \left[1 - a(k) \right]\left(1 - \varepsilon\right)^{b(k)} \tag{4.5}$$

where $\varepsilon = \sqrt{1 - \alpha_2^2}$ is a spectral width parameter and $\alpha(k)$ and $b(k)$ are best fitting parameters expressed as:

$$a(k) = 0.926 - 0.033k \;\; ; \qquad b(k) = 1.587k - 2.323 \tag{4.6}$$

This particularly simple expression was established based on observation of the data obtained from a rainflow analysis of simulated samples of some broad-band processes.

51

For a narrow-band process, $\alpha_2 = 1$ ($\varepsilon = 0$) and the expression correctly gives $\rho_{WL} = 1$. We note also that the previous formula assumes that the rainflow damage depends on just three spectral moments (i.e. λ_0, λ_2 and λ_4), through α_2 parameter.

This approach was applied to practical problems on wind and wave induced random loadings [Siddiqui and Ahmad 2001, Holmes 2002, Kukkanen and Mikkola 2004].

4.2.3. Approximations for the rainflow amplitude distribution

Some methods presented in the literature try to directly estimate the rainflow amplitude distribution $p_{RFC}(s)$, which is then used in Eq. (3.39) to compute fatigue damage under the linear rule.

In this approach, the main problem is that the true shape of the probability density function of rainflow amplitudes is generally unknown, as well as the set of spectral parameters that relate this distribution to the process power spectral density. Consequently, some kind of parametric shape must be assumed in advance and then calibrated through a best fitting procedure on data from extensive numerical simulations or experimental tests. For instance, densities used are often of Rayleigh, Exponential or Weibull type, or some kind of mixture [Wirsching and Sheata 1977, Bouyssy et al. 1993, Nagode and Faidiga 1998b].

On the other hand, the advantage of the knowledge of the amplitude distribution is twofold: first of all, it allows estimating the fatigue damage under the linear rule by simple integration as in Eq. (3.39). Furthermore, it allows treating rare events, as large cycles, by extrapolation of the amplitude distribution to large amplitude values.

4.2.3.1. Dirlik model (1985)

The most famous empirical formula for approximating the rainflow amplitude distribution is probably that proposed by Dirlik [Dirlik 1985], which uses a combination of an Exponential and two Rayleigh probability densities.

In the Dirlik model, the approximate closed-form expression for the probability density function of rainflow ranges r is [Bishop 1994, Halfpenny 1999, Haiba et al. 2002]:

$$p_{RFC}^{DK}(r) = \frac{1}{2(\lambda_0)^{1/2}} \left[\frac{D_1}{Q} e^{-\frac{Z}{Q}} + \frac{D_2 Z}{R^2} e^{-\frac{Z^2}{2R^2}} + D_3 Z e^{-\frac{Z^2}{2}} \right] \tag{4.7}$$

where:

$$Z = \frac{r}{2(\lambda_0)^{1/2}} = \frac{s}{(\lambda_0)^{1/2}} \tag{4.8}$$

is the normalised amplitude and:

$$x_m = \frac{\lambda_1}{\lambda_0}\left(\frac{\lambda_2}{\lambda_4}\right)^{1/2} \quad , \qquad D_1 = \frac{2(x_m - \alpha_2^2)}{1 + \alpha_2^2}$$

$$D_2 = \frac{1 - \alpha_2 - D_1 + D_1^2}{1 - R} \quad , \qquad D_3 = 1 - D_1 - D_2 \tag{4.9}$$

$$Q = \frac{1.25(\alpha_2 - D_3 - (D_2 R))}{D_1} \quad , \qquad R = \frac{\alpha_2 - x_m - D_1^2}{1 - \alpha_2 - D_1 + D_1^2}$$

are parameters resulting from a best fitting procedure over a large set of data from numerical simulations. It can be easily verified that $x_m = \alpha_1 \cdot \alpha_2$ and that $D_1/Q = 0.8$ (i.e. first coefficient in the distribution is constant).

The amplitude probability density, say $p_{RFC}^{DK}(s)$, follows from a simple variable transformation:

$$p_{RFC}^{DK}(s) = \frac{1}{(\lambda_0)^{1/2}}\left[\frac{D_1}{Q}e^{-\frac{Z}{Q}} + \frac{D_2 Z}{R^2}e^{-\frac{Z^2}{2R^2}} + D_3 Z e^{-\frac{Z^2}{2}}\right] \tag{4.10}$$

being as usual Z the normalised amplitude. It should be noted that, if compared to the Wirsching-Light model, this approach gives an amplitude distribution (and thus a rainflow damage) depending on just four spectral moments (i.e. λ_0, λ_1, λ_2 and λ_4), including in particular a dependence on λ_1 moment.

The rainflow damage intensity under the Palmgren-Miner rule is calculated by substituting $p_{RFC}^{DK}(s)$ density in Eq. (3.39):

$$\overline{D}_{RFC}^{DK} = \frac{v_p}{C}\lambda_0^{k/2}\left[D_1 Q^k \Gamma(1+k) + \left(\sqrt{2}\right)^k \Gamma\left(1 + \frac{k}{2}\right)\left(D_2 |R|^k + D_3\right)\right] \tag{4.11}$$

Finally, we compute the loading spectrum as in Eq. (3.35):

$$F^{DK}(s) = D_1 e^{-\frac{Z}{Q}} + D_2 e^{-\frac{Z^2}{2R^2}} + D_3 e^{-\frac{Z^2}{2}} \tag{4.12}$$

which, according to the definition of parameters, satisfies $F(s = 0) = 1$.

Several publications in the literature showed how the Dirlik method is far superior to other existing methods in estimating rainflow fatigue damage, see for example [Bouyssy et al. 1993, Bishop 1994, Halfpeny 1999]. However, we can notice how Dirlik method has some drawbacks. First of all, it was developed as a completely approximate approach, not supported by any kind of theoretical framework. Secondly, the proposed rainflow distribution does not account for mean value dependence, making so impossible a further extension to non-Gaussian problems (see Chapter 7 for more details).

4.2.3.2. Zhao-Baker model (1992)

Zhao and Baker used a similar concept, by assuming that amplitude probability distribution is a linear combination of one Weibull and one Rayleigh density [Zhao and Baker 1992]:

$$p_{RFC}^{ZB}(Z) = w \ a \, b \, Z^{b-1} \, e^{-a Z^b} + (1-w) \ Z \, e^{-\frac{Z^2}{2}} \qquad (4.13)$$

where Z is the normalised amplitude defined in Eq. (4.8), w is a weighting factor ($0 \le w \le 1$), and a, b are the Weibull parameters ($a > 0$, $b > 0$). Previous parameters, depending on spectral properties of process $X(t)$, are determined from simulations on a wide range of spectra, but are also supported by some theoretical arguments. Specifically, the weighting factor is:

$$w = \frac{1-\alpha_2}{1 - \sqrt{\dfrac{2}{\pi}} \, \Gamma\!\left(1+\dfrac{1}{b}\right) a^{-1/b}} \qquad (4.14)$$

while the other two parameters are:

$$a = 8 - 7\alpha_2 \quad , \qquad b = \begin{cases} 1.1 & \text{if } \alpha_2 < 0.9 \\ 1.1 + 9(\alpha_2 - 0.9) & \text{if } \alpha_2 \ge 0.9 \end{cases} \qquad (4.15)$$

For a narrow-band process, $\alpha_2 = 1$, which gives $a = 1$, $b = 2$ and $w = 0$, implying for the amplitudes a Rayleigh distribution, which is the exact distribution. However, according to the definition of a and b parameter given above, when $\alpha_2 \le 0.130$ it happens that $w > 1$, which is not correct; however, applications having so small values of α_2 are not so frequent in practice.

An alternative improved version of a parameter, which includes an additional functional relationship on $\alpha_{0.75}$ bandwidth parameter, exists. In fact, it was observed by simulations that, for small values of k (e.g. $k = 3$), rainflow damage is more closely correlated with other spectral properties than with α_2 [Lutes at. al 1984]. Specifically, the damage correction factor $\rho = \overline{D}_{RFC}/\overline{D}_{NB}$ was correlated with $\alpha_{0.75}$, for $k = 3$, by the following formula [Zhao and Baker 1992]:

$$\rho_{ZB}\big|_{k=3} = \begin{cases} -0.4154 + 1.392\,\alpha_{0.75} & \text{if } \alpha_{0.75} \ge 0.5 \\ 0.28 & \text{if } \alpha_{0.75} < 0.5 \end{cases} \qquad (4.16)$$

Then, a is calculated as $a = d^{-b}$, being d a solution of:

$$\Gamma\!\left(1+\frac{3}{b}\right)(1-\alpha_2)d^3 + 3\Gamma\!\left(1+\frac{1}{b}\right)(\rho_{ZB}\alpha_2 - 1)d + 3\sqrt{\frac{\pi}{2}}\,\alpha_2(1-\rho_{ZB}) = 0$$

$$(4.17)$$

In the case of a narrow-band process, $\alpha_{0.75} = 1$, so giving $\rho_{ZB} = 0.9766$, which is not coincident with the exact solution being expected (i.e. $\rho_{ZB} = 1$). Furthermore, by adopting this alternative definition, it can happen that $w < 0$ when considering particular values of $\alpha_{0.75}$ and α_2 (e.g. $\alpha_2 > 0.5$ and $\alpha_{0.75} < 0.65$). All details for calculating all parameters can be found in [Zhao and Baker 1992].

In the next, we will refer to the Zhao-Baker as the simplified version when w, a and b are given by Eqs. (4.14) and (4.15), to the improved version when a is given by Eqs. (4.16) and (4.17).

The amplitude density defined in Eq. (4.13) depends on the normalised amplitude; by a simple variable change, we can express it function of amplitude s, i.e.:

$$
p_{RFC}^{ZB}(s) = w\, \frac{a\,b}{\lambda_0^{1/2}} \left(\frac{s}{\lambda_0^{1/2}} \right)^{b-1} e^{-a\left(\frac{s}{\lambda_0^{1/2}} \right)^b} + (1-w)\, \frac{s}{\lambda_0}\, e^{-\frac{1}{2}\left(\frac{s}{\lambda_0^{1/2}} \right)^2} \tag{4.18}
$$

The rainflow damage intensity under the Palmgren-Miner rule is calculated by substituting $p_{RFC}^{ZB}(s)$ density as in Eq. (3.39):

$$
\overline{D}_{RFC}^{ZB} = \frac{v_p}{C}\, \lambda_0^{k/2} \left[w\, a^{-\frac{k}{b}}\, \Gamma\!\left(1 + \frac{k}{b} \right) + (1-w)\, 2^{\frac{k}{2}}\, \Gamma\!\left(1 + \frac{k}{2} \right) \right] \tag{4.19}
$$

Finally, we give the rainflow loading spectrum:

$$
F_{RFC}^{ZB}(s) = w\, e^{-a\left(\frac{s}{\lambda_0^{1/2}} \right)^b} + (1-w)\, e^{-\frac{s^2}{2\,\lambda_0^{1/2}}} \tag{4.20}
$$

4.2.3.3. Petrucci-Zuccarello method (2004)

A critical comparison of previously reviewed models highlights that the rainflow amplitude distribution may depend on the process spectral density through several spectral moments, as in Dirlik and Zhao-Baker models.

However, the true set of spectral moments involved is actually unknown and only reasonable hypotheses based on analysis of simulated results can be drawn. As an example, on the basis of results from numerical simulations, Petrucci and co-workers suggested a dependence on additional bandwidth parameters (four, at least), involving other (higher-order) spectral moments. More precisely, they proposed to take into, besides α_1 and α_2 for process $X(t)$, also corresponding parameters β_1 and β_2 for the derivative process $\dot{X}(t)$ (see Chapter 3) [Petrucci et al. 2000].

Unfortunately, they do not provide any theoretical justification that supports the validity of this hypothesis and furthermore an exhaustive theoretical solution including all mentioned parameters is not given, except an approximate method based on α_1 and α_2 parameters that will be illustrated here.

55

Including the mean value dependence into fatigue damage calculation, Eq. (3.39) for predicting the fatigue damage intensity modifies as:

$$\overline{D}_{\text{RFC}}^{\text{PZ}} = v_{\text{p}} \, C^{-1} \int_0^{+\infty} r_e^k \, p(r_e) \, dr_e = v_{\text{p}} \, C^{-1} \, \chi_k \tag{4.21}$$

where r_e is the equivalent range according to Goodman:

$$r_e = \frac{r}{1 - m/S_{\text{u}}} \tag{4.22}$$

being S_{u} the ultimate tensile stress of the material. The χ_k variable is k-th moment of the probability density function $p(r_e)$:

$$\chi_k = \int_0^{+\infty} r_e^k \, p(r_e) \, dr_e \tag{4.23}$$

Based on Eq. (4.23), the problem of rainflow fatigue damage prediction becomes the problem of the correct evaluation of the χ_k moment in terms of several spectral parameters. According to previous hypothesis, a dependence on just four bandwidth parameters (relative to $X(t)$ and $\dot{X}(t)$ processes) is assumed as:

$$\chi_k = \lambda_0^{k/2} \, f(\alpha_1, \alpha_2, \beta_1, \beta_2, \breve{R}) \tag{4.24}$$

where $\breve{R} = \sigma_{\max}/S_u$ is a dimensionless parameter related to the maximum stress $\sigma_{\max}$ of the process (this one usually assumed equal to three standard deviations).

As said before, a simplified dependence involving only α_1 and α_2 parameters is proposed by Petrucci and Zuccarello as a first approximation solution [Petrucci and Zuccarello 2004]:

$$\chi_k = \lambda_0^{k/2} \, \varphi(\alpha_1, \alpha_2, \breve{R}) \tag{4.25}$$

where $\varphi(\cdot)$ function is assumed as:

$$\varphi(\alpha_1, \alpha_2, \breve{R}) = e^{\Psi(\alpha_1, \alpha_2, k, \breve{R})} \tag{4.26}$$

and $\Psi(\cdot)$ is sought in polynomial form as:

$$\Psi(\alpha_1, \alpha_2, k, \breve{R}) = \frac{(\Psi_2 - \Psi_1)}{6}(k - 3) + \Psi_1 + \\ + \left[\frac{2}{9}(\Psi_4 - \Psi_3 - \Psi_2 + \Psi_1)(k - 3) + \frac{4}{3}(\Psi_3 - \Psi_1)\right](\breve{R} - 0.15) \tag{4.27}$$

$$\Psi_1 = -1.994 - 9.381\,\alpha_2 + 18.349\,\alpha_1 + 15.261\,\alpha_1\alpha_2 - 1.483\,\alpha_2^2 - 15.402\,\alpha_1^2$$
$$\Psi_2 = 8.229 - 26.510\,\alpha_2 + 21.522\,\alpha_1 + 27.748\,\alpha_1\alpha_2 + 4.338\,\alpha_2^2 - 20.026\,\alpha_1^2$$
$$\Psi_3 = -0.946 - 8.025\,\alpha_2 + 15.692\,\alpha_1 + 11.867\,\alpha_1\alpha_2 + 0.382\,\alpha_2^2 - 13.198\,\alpha_1^2$$
$$\Psi_4 = 8.780 - 26.058\,\alpha_2 + 21.628\,\alpha_1 + 26.487\,\alpha_1\alpha_2 + 5.379\,\alpha_2^2 - 19.967\,\alpha_1^2$$

$$(4.28)$$

Consequently, knowledge of the material (static and fatigue) strength properties k and S_u, and of the process spectral properties as λ_0, α_1 and α_2, Eqs. (4.21)-(4.28) allow one to compute the rainflow damage intensity.

4.2.3.4. Tovo-Benasciutti method [Tovo 2002, Benasciutti and Tovo 2005a, 2006a]

In this section we analyse in detail an alternative method for estimating the rainflow cycle distribution (and the related fatigue damage under the linear rule) in a Gaussian broad-band processes.

Let $x(t)$, $0 \le t \le T$ be a time history belonging to a Gaussian random process $X(t)$. From Chapter 3 we know that:

$$\overline{D}_{RC} \le \overline{D}_{RFC} \le \overline{D}^+ = \overline{D}_{NB} \tag{4.29}$$

This equation states that, under the linear damage accumulation rule, the rainflow damage always bounds the damage from the range count, and that an upper damage value, $\overline{D}^+$, exists, which bounds damage computed by any crossing-consistent counting method (see Chapter 3).

Rychlik proved that, in Gaussian processes, the upper bound coincides with the damage given by the narrow-band approximation, Eq. (4.3) [Rychlik 1993a]. Furthermore, it has been recently pointed out that the narrow-band damage also equals the damage calculated under the linear damage rule for the level-crossing count [Tovo 2002].

Precisely, in the level-crossing count all positive peaks, reduced by the number of positive valleys at the same level, are paired with the lowest available valley, which is symmetric in a symmetric process, to form damaging cycles with non-zero amplitude. The remaining peaks and valleys at same level are paired together to form zero-amplitude (non-damaging) cycles. This leads to the following distribution of level-crossing counted cycles as a function of peak and valley levels:

$$h_{LCC}(u,v) = \begin{cases} \left[p_p(u) - p_v(u) \right]\delta(u+v) + p_v(u)\delta(u-v) & \text{if } u > 0 \\ p_p(u)\delta(u-v) & \text{if } u \le 0 \end{cases} \tag{4.30}$$

where $\delta(\cdot)$ is the Dirac delta function and $p_p(u)$ and $p_v(u)$ are the peak and valley distributions, respectively. The component related to $\delta(u-v)$ represents zero-amplitude cycles, has no damaging effect and may be neglected in practical applications; instead, the component related to $\delta(u+v)$ represents non-zero amplitude (damaging) cycles. In the following, we shall use indifferently the notation $h_{\text{LCC}}(u,v)$ or $h_{\text{NB}}(u,v)$ for indicating the cycle distribution of the narrow-band approximation.

The distribution $h_{\text{LCC}}(u,v)$ as given in Eq. (4.30) is quite general and it is valid for both Gaussian and non-Gaussian symmetric loadings, if proper peak and valley distributions are used. In the case of Gaussian processes, whose peak and valley distributions are known, the density $h_{\text{LCC}}(u,v)$ is a solution of the integral equation Eq. (3.33) in Chapter 3, confirming that the level-crossing count is a complete procedure (hence, $v_a = v_p$).

The distribution of amplitude and mean value associated to $h_{\text{LCC}}(u,v)$ is:

$$p_{a,m}^{\text{LCC}}(s,m) = \begin{cases} \left[p_p(s) - p_v(s) \right] \delta(m) + p_v(m)\delta(s) & \text{if } s+m>0 \\ p_p(m)\delta(s) & \text{if } s+m\le 0 \end{cases} \qquad (4.31)$$

and it is depicted in Figure 4.2(a).

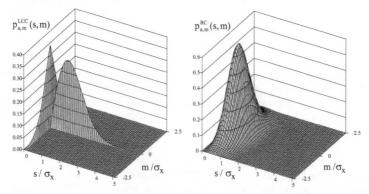

Figure 4.2: Probability density distributions of counted cycles ($p_{a,m}^{\text{LCC}}$ representation is only qualitative, since Dirac delta function can not be plotted). (Reprinted from [Tovo 2002], with permission from Elsevier)

The marginal amplitude distribution is:

$$p_a^{\text{LCC}}(s) = [p_p(s) - p_v(s)] \; = \; \alpha_2 \frac{s}{\lambda_0} e^{-\frac{s^2}{2\lambda_0}} \qquad (4.32)$$

and it is a Rayleigh density that, substituted into Eq. (3.39), with v_p given by Eq. (3.7), gives the damage intensity $\overline{D}_{\text{NB}}$ as in Eq. (4.3), i.e. the narrow-band approximation. This proves that, in the case of a Gaussian process, the joint density $h_{\text{LCC}}(u,v)$ is the expression of the distribution of both the level-crossing count and that the upper bound of any crossing-consistent cycle count. In fact, integration of the $h_{LCC}(u,v)$ distribution as in Eq. (3.34) gives its cumulative counting intensity:

$$\mu_{LCC}(u,v) = v(u)\,\mathbf{I}(u+v) + v(v)\,\mathbf{I}(-(u+v))$$ (4.33)

where $\mathbf{I}(\cdot)$ is an indicator function ($\mathbf{I}(x)=1$ if $x \geq 0$, elsewhere $\mathbf{I}(x)=0$) and $v(u)$ is the level-crossing spectrum, Eq. (3.17). Equation above clearly equals the $\mu^+(u,v)$ count intensity defined by Eq. (3.27), showing that the level-crossing is a crossing-consistent count.

For the lower bound of the rainflow damage (i.e. the range count damage), at present no exact analytical expression is known, then we can adopt the approximate result proposed in [Madsen et al. 1986]:

$$\overline{D}_{RC} \cong v_P\,C^{-1}\left(\sqrt{2\lambda_0}\,\alpha_2\right)^k \Gamma\!\left(1+\frac{k}{2}\right) = \overline{D}_{NB}\,\alpha_2^{k-1}$$ (4.34)

The above formula was obtained by studying the double envelope of a random process and by approximating the result of the range count by means of the amplitude of the envelope process (further details on differences between this definition and the actual range count are given in [Madsen at. al. 1986]).

It is noteworthy that we also know explicitly the underlying cycle distribution $h_{RC}(u,v)$, which gives the approximate range count damage, $\overline{D}_{RC}$, under the assumption of the linear damage accumulation rule.

Precisely, let us consider the following particular joint density proposed in [Tovo 2002]:

$$h_{RC}(u,v) = \frac{1}{2\sqrt{2\pi}\,\lambda_0\,\alpha_2^2}\,e^{-\frac{u^2+v^2}{4\lambda_0(1-\alpha_2^2)}}\,e^{-\frac{(u-v)^2}{4\lambda_0(1-\alpha_2^2)}\frac{1-2\alpha_2^2}{2\alpha_2^2}}\left[\frac{u-v}{\sqrt{4\lambda_0(1-\alpha_2^2)}}\right]$$ (4.35)

The corresponding joint distribution of amplitudes and mean values, $p_{a,m}^{RC}(s,m)$, and the marginal amplitude distribution, $p_a^{RC}(s)$, are:

$$p_{a,m}^{RC}(s,m) = \frac{1}{\sqrt{2\pi\,\lambda_0(1-\alpha_2^2)}}\,e^{-\frac{m^2}{2\lambda_0(1-\alpha_2^2)}} \cdot \frac{s}{\lambda_0\,\alpha_2^2}\,e^{-\frac{s^2}{2\alpha_2^2\lambda_0}}$$ (4.36)

$$p_a^{RC}(s) = \frac{s}{\lambda_0\,\alpha_2^2}\,e^{-\frac{s^2}{2\alpha_2^2\lambda_0}}$$ (4.37)

The probability density $p_{a,m}^{RC}(s,m)$ is depicted in Figure 4.2(b).

The important point here is that cycle distribution $h_{RC}(u,v)$ gives the range counting damage $\overline{D}_{RC}$ under the Palmgren-Miner damage rule, i.e. when we substitute its related marginal amplitude distribution $p_a^{RC}(s)$ into Eq. (3.39). This means that the distribution represented by Eqs. (4.35)-(4.37) causes the same damage, as that proposed by Madsen et al. as an approximation of the range count damage. Obviously, this does not mean that these densities are the cycle distributions resulting from the range count, but it is reasonable that they do cause a damage close to the lower bound of the rainflow counting damage.

Further researches have shown that $h_{RC}(u,v)$ density coincides with an approximate function quantifying the transition probability between adjacent extremes in a Gaussian process, independently proposed by Sjöström and Kowalewski [Sjöström 1961, Kowalewski 1966] (Kowalewski's formula can be also found in [Bishop and Sherrat 1990, Stichel and Knothe 1998]). In other words, Sjöström-Kowalewski' joint density, $h_{RC}(u,v)$, can be viewed as an approximate distribution for cycles identified by the range count, i.e. cycles constructed by pairing adjacent local extremes (e.g. a maximum and the following minimum). Other references related to this distribution are referred to Butler (1961) and to Cartwright and Longuet-Higgins (1956), as indicated in [Tunna 1986].

Several properties of the $h_{RC}(u,v)$ density are of interest. First of all, the count intensity calculated according to Eq. (3.34), after some manipulations, yields:

$$\mu_{RC}(u,v) = v_0 \left\{ e^{-\frac{v^2}{2\lambda_0}} \left[1 - \Phi\left(\frac{u - v(1-2\alpha_2^2)}{2\,\alpha_2\,\sqrt{\lambda_0\,(1-\alpha_2^2)}} \right) \right] + \right.$$

$$\left. + e^{-\frac{u^2}{2\lambda_0}} \Phi\left(\frac{v - u(1-2\alpha_2^2)}{2\,\alpha_2\,\sqrt{\lambda_0\,(1-\alpha_2^2)}} \right) \right\}$$

(4.38)

where v_0 is the mean upcrossing rate. It is straightforward to prove that, for $u = v$, previous expression converts into the upcrossing formula for a Gaussian process (Rice's formula), i.e.:

$$\mu_{RC}(u,u) = v_0 \left\{ e^{-\frac{u^2}{2\lambda_0}} \left[1 - \Phi\left(\frac{\alpha_2\,u}{\sqrt{\lambda_0\,(1-\alpha_2^2)}} \right) \right] + e^{-\frac{u^2}{2\lambda_0}} \Phi\left(\frac{\alpha_2\,u}{\sqrt{\lambda_0\,(1-\alpha_2^2)}} \right) \right\}$$

$$= v_0\, e^{-\frac{u^2}{2\lambda_0}} = v(u)$$

(4.39)

which confirms that the counting method with $h_{RC}(u,v)$ distribution is crossing-consistent. In a Gaussian process, where peak and valley distributions are known, it is possible to verify by integration that $h_{RC}(u,v)$ density is a solution of Eq. (3.33) in Chapter 3, i.e. it is the cycle distribution of a counting method which is complete. Figure 4.3 shows the level curves of $\mu_{LCC}(u,v)$ and $\mu_{RC}(u,v)$ for a Gaussian process having $\lambda_0 = \lambda_2 = 1$ and $\lambda_4 = 4$ ($\alpha_2 = 0.5$).

Secondly, careful consideration of Eq. (4.36) reveals that $p_{a,m}^{RC}(s,m)$ density actually represents independently distributed amplitude and mean value random variables, the former having a Rayleigh and the latter a Gaussian probability density function, with variance $\sigma_s^2 = \sqrt{\lambda_0 (1-\alpha_2^2)}$ and $\sigma_m^2 = \lambda_0 \alpha_2^2$, respectively. This can be condensed using the following symbolic notation:

$$p_{a,m}^{RC}(s,m) \overset{D}{=} \sqrt{\lambda_0 (1-\alpha_2^2)}\, U \;\cdot\; \lambda_0^{1/2}\alpha_2\, R \tag{4.40}$$

where U is a standard normal variable and R a standard Rayleigh variable (U independent of R), and where $\overset{D}{=}$ denotes that two variables have same distribution. Note that $\sigma_s^2 + \sigma_m^2 = 1$. It can be rigorously proved that, in a Gaussian process, exact independence between amplitudes and mean values of range counted cycles would imply that amplitudes have a Rayleigh distribution and that mean values are Gaussian [Lindgren and Broberg 2003].

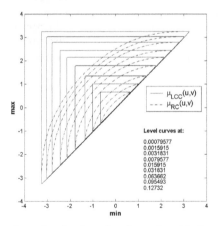

Figure 4.3: Comparison of $\mu_{LCC}(u,v)$ and $\mu_{RC}(u,v)$ count intensities for a Gaussian process with $\lambda_0 = \lambda_2 = 1$ and $\lambda_4 = 4$ ($\alpha_2 = 0.5$).

Some additional comments are now of interest. Systematic analysis of results from extensive numerical simulations has confirmed the correctness of the hypothesis of a

Gaussian density (obviously with a zero mean value) for the distribution of mean values of range counted cycles. At the same time, the hypothesis of a Rayleigh distribution for amplitudes has been found correct only for a restrict class of processes, namely for broad-band processes whose derivative is narrow-band ($\beta_1 \approx 1$; $\beta_2 \approx 1$). In both cases, the variance of the distribution is calculated as a polynomial, depending on α_1 (or q_X) and α_2 variables [Petrucci and Zuccarello 1999].

Consequently, for Gaussian broad-band processes with a narrow-band derivative process, reasonably accurate fatigue life predictions are made assuming the hypothesis of independently distributed amplitudes and mean values (errors are less than 10 per cent). In the general case (e.g. for whatever broad-band process), the hypothesis of independence (on which the Sjöström-Kowalewski' density is based) leads generally to large errors [Petrucci and Zuccarello 1999].

For example, numerical simulations have shown that mean values are almost independent of amplitudes only in a loading with a spectral density having a rectangular shape [Lindgren and Broberg 2003]. Anyway, we must always remember that the independence assumption never holds for any Gaussian process with non-zero bandwidth, as demonstrated in [Lindgren 1970].

Previous discussion underlined that the distribution of range counted cycle is strictly related to transition probability between adjacent extremes. This fact enables us for example to construct a Markov matrix after suitable normalisation of $h_{RC}(u,v)$ density. Otherwise, the transition matrix can be estimated directly from measured or simulated data, or alternatively by numerical procedures. Two examples are mentioned here. The first regards a numerical-based approach for determining the one-step transition matrix (and thus the max-min cycle distribution) proposed in [Krenk and Gluver 1989]. The interesting fact is that the transition probability of small ranges shows a dependence on β_2 parameter, i.e. the irregularity factor of the derivative process. The second is a complete numerically-based method (available in WAFO toolbox) developed for the determination of the range count cycle distribution; the method, although numerical, is exact in the sense that distribution becomes asymptotically exact when the integration grid increases [Lindgren and Broberg 2003].

We turn now to the evaluation of the rainflow fatigue damage. As stated by Eq. (4.29), the rainflow damage is always placed in-between previously defined bounds, namely $\overline{D}_{NB}$ and $\overline{D}_{RC}$, which for a given process (i.e. for a given spectral density) are fixed quantities, see Eqs. (4.3) and (4.34).

Thus, the problem of finding the rainflow damage intensity $\overline{D}_{RFC}$ becomes the problem of finding the proper intermediate point between these bounds. A linear combination is then suggested:

$$\overline{D}_{RFC} = b\,\overline{D}_{NB} + (1-b)\,\overline{D}_{RC} \qquad (4.41)$$

in which the b weighting factor is assumed to depend on the spectral and bandwidth parameters of the power spectral density of the random process.

On the basis of Eq. (4.29), we can expect that a relation similar to Eq. (4.41) is also true for cycle distributions. Therefore, we can estimate the distribution of rainflow counted cycles, $h_{RFC}(u,v)$ (or equivalently its related cumulative distributions, $H_{RFC}(u,v)$ or $\mu_{RFC}(u,v)$) by using an analogous linear combination. For example, the rainflow joint density function is estimated as:

$$h_{RFC}(u,v) = b\, h_{LCC}(u,v) + (1-b)\, h_{RC}(u,v) \tag{4.42}$$

being $h_{LCC}(u,v)$ and $h_{RC}(u,v)$ the cycle distributions for level-crossing (i.e. the narrow-band approximation) and range counts. Since the distribution of range counted cycles is computed by an approximate formula, the Eq. (4.42) holds only as a first approximation.

Similarly, we can estimate the probability density of rainflow cycles as:

$$p_a^{RFC}(s) = b\, p_a^{LCC}(s) + (1-b)\, p_a^{RC}(s) \tag{4.43}$$

being $p_a^{LCC}(s)$ and $p_a^{RC}(s)$ the amplitude probability densities for the level-crossing and range count. The advantage of formula (4.42) is that it estimates the rainflow cycle distribution in terms of peak and valley levels, then, it allows us to further develop the method to non-Gaussian processes.

Based on Eq. (4.34), the expression in Eq. (4.41) can also be written as:

$$\overline{D}_{RFC} = \left[b + (1-b)\,\alpha_2^{k-1} \right] \overline{D}_{NB} = \rho_{TB}\, \overline{D}_{NB} \tag{4.44}$$

in which λ_{TB}, in analogy with Eq. (4.4), can be interpreted as a correction of the narrow-band approximation. However, the main difference here in respect to the Wirsching-Light method, is that a more complete theoretical framework is behind the definition of λ_{TB} index.

Until now, no exhaustive theoretical information concerning b parameter and its dependence on process spectral density is available. However, some general properties can be mentioned here. In a narrow-band process, $\alpha_2 = 1$ and then $\overline{D}_{RFC}$ equals $\overline{D}_{NB}$, whatever value b may have, which seems correct. Furthermore, when $k = 1$, Eq. (4.44) predicts $\overline{D}_{RFC} = \overline{D}_{NB}$, which is true [Rychlik 1993a, Lutes at al. 1984]. Finally, Eq. (4.42) is also applicable in the case of an irregular process having $\alpha_2 = 0$, where it reduces to $\overline{D}_{RFC} = b\, \overline{D}_{NB}$.

More specifically, since we do not know the exact correlation relating b to spectral parameters, we must rely only on approximate formulas based on reasonable assumptions and calibrated on numerical simulations; for example, the following formula [Tovo 2002]:

$$b_{app}^{Tov} = \min\left\{ \frac{\alpha_1 - \alpha_2}{1 - \alpha_1}, 1 \right\} \tag{4.45}$$

implicitly assumes that the rainflow damage depends on just four spectral moments (i.e. λ_0, λ_1, λ_2 and λ_4) through α_1 and α_2 bandwidth parameters (compared to other methods, once again a further dependence on λ_1 moment is introduced, as in Dirlik method).

In the next section, results from numerical simulations will be used to verify the accuracy of previous expression and to propose a modified improved approximation, which still involve only α_1 and α_2 parameters. As discussed above, a more complex dependence including other spectral parameters (as the bandwidth parameters relative to the derivative process $\dot{X}(t)$) could yet exist.

4.2.4. The empirical-$\alpha_{0.75}$ method

As observed in previous sections, a possible correlation of rainflow fatigue damage on some particular bandwidth parameters has been investigated and a dependence on $\alpha_{0.75}$ was suggested in [Lutes et al. 1984].

On a pure empirical basis, one can argue that the damage correction factor $\rho = \overline{D}_{RFC}/\overline{D}_{NB}$ is only a function of $\alpha_{0.75}$ bandwidth parameter, and that it is independent of S-N slope k. The following simple formulation is then suggested:

$$\overline{D}_{RFC} = \rho_{0.75}\,\overline{D}_{NB} \qquad \rho_{0.75} = \alpha_{0.75}^2 \tag{4.46}$$

where $\rho_{0.75}$ indicates the damage correction factor for this method. The above expression, even if approximate and lacking of any sort of theoretical motivation, was shown to agree fairly well with data from simulation [Benasciutti and Tovo 2004a, 2004b, 2006a] and can be taken as a first approximation of the rainflow damage.

4.3. NUMERICAL SIMULATIONS

Numerical simulations aim to validate the accuracy of spectral methods in estimating, in Gaussian broad-band processes, both the rainflow cycles distribution and the rainflow fatigue damage under the Palmgren-Miner law. Secondly, they also aim to investigate the true set of spectral parameters that correlate the power spectral density of random process to the statistical distribution of rainflow counted cycles. Results of this section will be mainly focused on the properties of b index, defined by Eq. (4.42).

As discussed before, the true set of spectral moments controlling the rainflow cycle distribution is actually not known and we can only rely on reasonable hypotheses based on simulations.

The main assumption is that the distribution of rainflow cycles mainly depends on α_1 and α_2 bandwidth parameters; thus, the need of different spectral densities having the same α_1 and α_2 pair (or, alternatively, different bandwidth indexes from the same spectrum geometry) is assumed as the guideline of our simulations.

Several stochastic processes were numerically simulated by assuming different shapes of the spectral density; namely, various one-sided spectral densities, $W_X(\omega)$

were considered, having simple geometries like those depicted in Figure 4.4, e.g. constant, linear, double symmetrical or anti-symmetric parabolic shape.

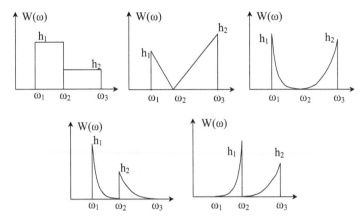

Figure 4.4: Spectral densities used in numerical simulations.

All spectra have the same variance (i.e. λ_0 moment) and common values for ω_1 and ω_3 frequencies (see Table 4.2), whereas ω_2 can move arbitrarily inside them. For a given spectrum and ω_2 value, a well-defined set of α_1 and α_2 bandwidth parameters is univocally established by selecting a proper (h_1, h_2) pair.

Table 4.2: Common parameters in numerically simulated processes.

$\omega_1 = 2\pi$	$\omega_3 = 10^5 \omega_1$	$\lambda_0 = \sigma_X^2 = 10^4$

The complete set of numerical simulations focused on α_2 values of 0.1, 0.3, 0.5 and 0.7, with α_1 taking specified increasing values between 0.1 and 0.9. For a given couple of these two indexes, the choice of different spectral density shapes makes possible for β_1 parameter (i.e. the irregularity factor of the derivative process $\dot{X}(t)$) to range from 0.500 to 0.950. Figure 4.5 gives some examples of parts of simulated processes with different combinations of bandwidth parameters.

In each simulation test (i.e. given a spectral density) a time history is generated in the time-domain and then the traditional time-domain analysis is performed: first, cycles are counted by means of the rainflow count, then the fatigue damage is computed under the linear damage accumulation law. In fatigue damage computations, we assume two values for the slope k of the S-N relation (i.e. $k = 3$ and $k = 5$) and a reference strength C equal to unity.

Results from time-domain calculations (i.e. cycle distribution and fatigue damage) are compared with predictions made in the frequency-domain by all different spectral methods reviewed in previous sections. Precisely, five methods have been considered: the Wirsching-Light (WL), Zhao-Baker (ZB), Dirlik (DK), the Tovo-Benasciutti (TB) and the empirical-$\alpha_{0.75}$ method.

Although the approximate formula for b weighting factor given in Eq. (4.45) has been shown to be fairly accurate [Tovo 2002], our intention is to find an improved version of that formula. In order to do this, we have to understand the intrinsic relationship relating b coefficient to α_1 and α_2 bandwidth parameters, therefore we need to express Eq. (4.41) as a function of quantities resulting in simulations.

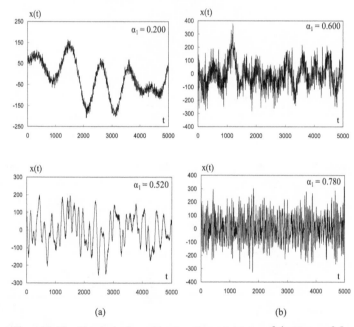

(a)   (b)

Figure 4.5: Time histories having various α_1 values and: (a) $\alpha_2 = 0.1$; (b) $\alpha_2 = 0.5$.

Time-domain calculation on a given simulated time history provides a fatigue damage value, say $\hat{D}_{RFC}$, which can be taken as an estimate of the expected rainflow damage. Thus, inverting Eq. (4.41) and substituting this damage value gives an estimate of b factor as:

$$\hat{b} = \frac{\hat{D}_{RFC} - \overline{D}_{RC}}{\overline{D}_{NB} - \overline{D}_{RC}} \tag{4.47}$$

being $\overline{D}_{NB}$ and $\overline{D}_{RC}$ the damage intensities calculated by Eqs. (4.3) and (4.34) which are only functions of the process spectral density.

Repeating such calculation for all simulation tests (i.e. for all spectral densities) provides a set of $\hat{b}$ values for different α_1 and α_2 pairs; all such simulation results are shown as marked points in Figure 4.6.

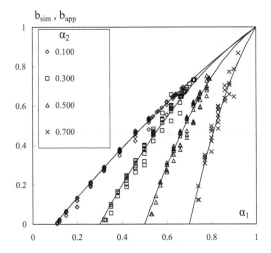

Figure 4.6: Comparison between $\hat{b}$ values in numerical simulations and approximate analytical prediction, Eq.(4.48).

Since, as stated by Eq. (4.41), the bandwidth parameters control rainflow damage through b coefficient, Figure 4.6 clearly highlights a strong dependence of damage on α_1 index, even for a constant α_2 value. Consequently, we may presume to have to consider also α_1 index in damage estimation formulas (as suggested by DK and the TB method). Anyway, the relative spread of simulated results observed in Figure 4.6 (associated to different β_1 values) can not be completely disregarded, meaning that a slight variation on damage, also for constant values of α_1 and α_2 parameters, can sometimes be observed.

Even if conscious of a possible additional relationship related to β_1 index, our opinion is to focus on a functional relationship still involving α_1 and α_2 parameters (as already proposed in Eq. (4.45)). A closed-form of such relation is given in analytical form by the following expression (see Appendix B):

$$b_{\text{app}} = \frac{(\alpha_1 - \alpha_2)\left[1.112\left(1 + \alpha_1\alpha_2 - (\alpha_1 + \alpha_2)\right)e^{2.11\alpha_2} + (\alpha_1 - \alpha_2)\right]}{(\alpha_2 - 1)^2} \quad (4.48)$$

which, added in Figure 4.6 as a continuous line, is shown to exhibit a satisfactory agreement with all numerical results.

Nevertheless, a possible dependence of the rainflow damage on $\alpha_{0.75}$ bandwidth parameter may exist. In Figure 4.7 we show the correlation existing between the expression $\rho_{0.75} = \alpha_{0.75}^2$ (continuous line) and the results from simulations (dots). As can be seen, all data are very close to the proposed expression, except for some points, independently of the S-N slope k.

At this point, we can compare the results from different spectral methods. First of all, we concentrate on the distribution of rainflow cycles, and then on the fatigue damage. The TB method is applied by using the b index given in Eq. (4.48).

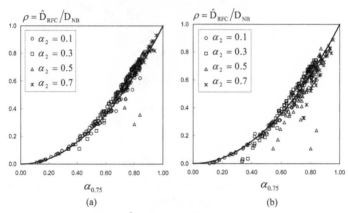

Figure 4.7: The expression $\rho_{0.75} = \alpha_{0.75}^2$ (continuous line) is compared with simulation results (dots). S-N slope (a) $k = 3$ and (b) $k = 5$.

Figure 4.8 and Figure 4.9 report an example of the comparison between the loading spectrum $\hat{F}_{RFC}(s)$ obtained from the time-domain rainflow analysis and the expected loading spectra provided by (improved) ZB, DK and TB methods, for a set of simulated time histories having α_2 equal to 0.1 and 0.5 respectively, and different values of α_1. Note that both the NB and WL approaches, besides the empirical-$\alpha_{0.75}$ method, can not be considered, since they do not supply any information about the rainflow cycle distribution.

As can be seen, a good agreement between the expected and the sample fatigue spectra is generally observed, even for the lowest value of α_2. Nevertheless, our calculations highlighted some drawbacks for the ZB technique. First of all, the simplified version could not be applied to all that processes with $\alpha_2 = 0.1$, resulting in a weight $w > 1$. On the other hand, also the improved version could give sometimes negative values for weight w (e.g. $\alpha_2 > 0.5$ and $\alpha_{0.75} < 0.65$). In addition, the improved version, providing results comparable to the simplified one, is often not as accurate as other methods, see for example Figure 7(a).

68

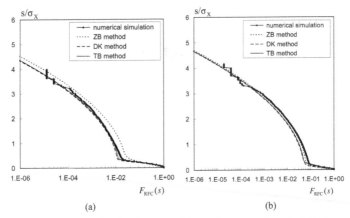

Figure 4.8: Comparison of rainflow loading spectra. The sample spectrum from simulation is compared with estimation from the DK, ZB (improved version) and the TB methods. Processes have $\alpha_2 = 0.100$ and (a) $\alpha_1 = 0.200$; (b) $\alpha_1 = 0.520$.

Results similar to those presented in Figure 4.8 and Figure 4.9 are observed for other processes, having different combinations of α_1 and α_2 bandwidth parameters.

For what concerns fatigue damage calculations, results are presented in Figure 4.10 through Figure 4.13. In Figure 4.10, the rainflow and the range count damage values given by time-domain analysis on simulated time histories are shown together with the damage estimations $\overline{D}_{NB}$ and $\overline{D}_{RC}$, i.e. the NB approximation and the approximate range-mean damage. Damage values are suitably normalised, so to give a (normalised) damage/cycle, i.e. $\overline{D}/(C \nu_p \sigma_X^k)$. Therefore, according to Eqs. (4.3) and (4.34), both the normalised $\overline{D}_{NB}$ and $\overline{D}_{RC}$ damage values will depend only on α_2 (which in each graph is constant).

Figure 4.10 confirms us that the fundamental relation, Eq. (4.29), is verified independently of α_2: the time-domain rainflow damage actually falls inside the two damage bounds and it also seems highly dependent on α_1 index, even for a fixed value of α_2, thus suggesting a possible correlation between the rainflow damage and λ_1 spectral moment, besides the dependence on λ_0, λ_2 and λ_4 spectral moments already expressed through α_2 index.

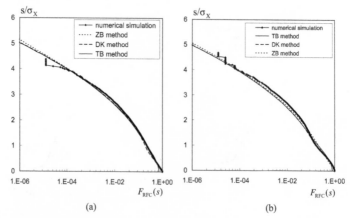

(a)   (b)

Figure 4.9: Comparison of rainflow loading spectra. The sample spectrum from simulation is compared with estimation from the DK, ZB (improved version) and the TB methods. Processes have $\alpha_2 = 0.500$ and (a) $\alpha_1 = 0.780$; (b) $\alpha_1 = 0.660$.

It is our opinion that improved damage estimations should adequately account for this correlation, either directly on λ_1 or through α_1 parameter (as done for example in both DK and TB methods). However, due to the complicate connexions existing amongst all spectral parameters (see Appendix A), a further more complex relationship could exist (e.g. involving a dependence on $\lambda_{0.75}$ and $\lambda_{1.5}$, as in the improved ZB method and the empirical-$\alpha_{0.75}$ method).

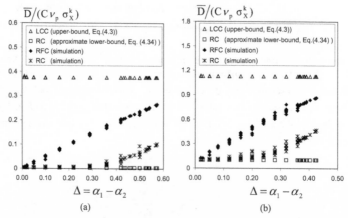

(a)   (b)

Figure 4.10: Damage value comparison. Analytical damage bounds of rainflow damage, Eqs. (4.3) and (4.34), together with the time-domain rainflow and range counting damage values calculated from simulated time histories having: (a) $\alpha_2 = 0.100$; (b) $\alpha_2 = 0.300$.

It is important to note that, as shown in Figure 4.10, also the range count damage given by simulations greatly depends on α_1. More precisely, the accuracy of the approximate range count damage, Eq. (4.34), is shown to decrease with $\Delta = \alpha_1 - \alpha_2$. Thus, the use of Eq. (4.34) for calculating the rainflow damage computation, as in Eq. (4.41), seems rather questionable. However, it should be noted in Figure 4.10 that damage value of Eq. (4.34) is always less than the actual range count damage from simulation, i.e. it is below the actual lower bound of rainflow damage. Furthermore, when Eq. (4.34) looses its validity (i.e. for high $\Delta = \alpha_1 - \alpha_2$ values), b_{app} coefficients calculated by either Eq. (4.45) or Eq. (4.48) tends to unity, see Figure 4.11, which so makes the contribution of $\overline{D}_{RC}$ damage less important in calculating the rainflow damage $\overline{D}_{RFC}$ in Eq. (4.41).

Anyway, on the basis of Figure 4.10 one can assert that Eqs. (4.3) and (4.34) for the two analytical damage bounds are very useful tools for a first approximation of the rainflow damage, except in situations in which α_2 and k assume too low and too high values, respectively (in fact, the spread between the two bounds is about α_2^{k-1}).

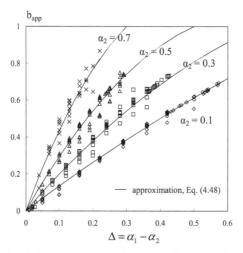

Figure 4.11: This figure plots results of Figure 4.6 as a function of $\Delta = \alpha_1 - \alpha_2$.

Finally, Figure 4.12 and Figure 4.13 qualitatively show an overall comparison of damage predictions by all spectral methods (results from the WL and the empirical-$\alpha_{0.75}$ method are then included). Results concerning rainflow fatigue damage from Gaussian simulation for each of the spectral densities types studied, having α_2 equal to 0.1, 0.3, 0.5 and 0.7.

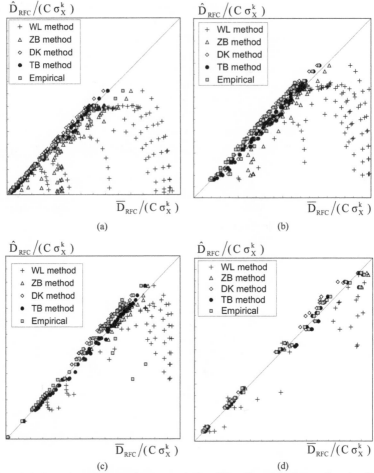

Figure 4.12: Comparison of damages for processes having different irregularity factors. Slope $k = 3$. (a) $\alpha_2 = 0.1$; (b) $\alpha_2 = 0.3$; (c) $\alpha_2 = 0.5$; (d) $\alpha_2 = 0.7$.

In each figure, the abscissa of the data points is the expected rainflow damage intensity as estimated by a spectral method, and the ordinate is the damage intensity as calculated by the rainflow analysis of the simulated data.

Thus, perfect correspondence between a spectral method and a simulation is indicated by data lying on the straight line, representing the bisector of the damage plane. Any deviations of the data from this line indicate inaccuracies in the spectral methods. Damage values are all normalised to the fatigue strength C and to σ_X^k, being constant for all spectral densities analysed.

72

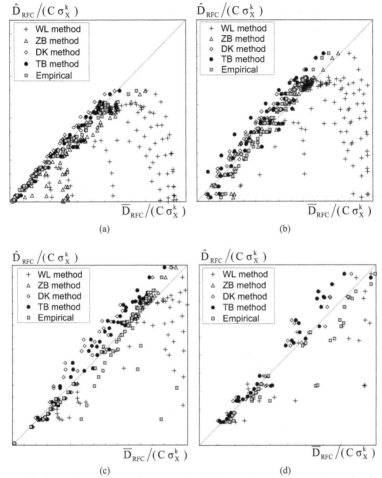

Figure 4.13: Comparison of damages for processes having different irregularity factors. Slope $k = 5$. (a) $\alpha_2 = 0.1$; (b) $\alpha_2 = 0.3$; (c) $\alpha_2 = 0.5$; (d) $\alpha_2 = 0.7$.

A simple way to quantify the accuracy of the estimation of a given spectral method is to introduce an error index EI, defined as the root mean square error of the damage estimation (note that other indexes can be used [Braccesi et al. 2005]):

$$EI = \sqrt{\frac{1}{n}\sum_i \left[\log_{10}\left(\frac{\overline{D}^*_{\mathrm{RFC},i}}{\hat{D}_{\mathrm{RFC},i}} \right) \right]^2} \qquad (4.49)$$

where the rainflow damage $\hat{D}_{\text{RFC},i}$ in the i-th simulation is from time-domain rainflow analysis, while damage $\overline{D}^*_{\text{RFC},i}$ as predicted by a given spectral method; the sum is extended over the entire set of simulation results, including both S-N slopes $k = 3$ and $k = 5$. The error indexes for all methods are summarised in the last column of Table 4.3.

Table 4.3: Comparison of estimations from the spectral methods considered in numerical simulations Applicability refers to the set of numerical simulations used in the present study.

Spectral method	Applicability	Accuracy
	% in our simulations	Error index [a]
Narrow-band approximation	100%	0.541
Wirsching -Light	100%	0.472
Zhao-Baker (simplified)	$\alpha_2 > 0.13$ 68%	0.253
Zhao-Baker (improved)	83%	0.297
Dirlik	100%	0.084
TB	100%	0.090
Empirical- $\alpha_{0.75}$	100%	0.105

[a] Error index: see Eq. (4.49)

Results presented in all figures highlight how the WL approach, even if quite simple if compared to other existing methods, tends to systematically give too conservative predictions in respect to simulated damage. The accuracy of the prediction increases when α_2 tends to unity, i.e. when the process tends to be narrow-band, which seems plausible. Thus, all the more reason that the narrow-band approximation can not be expected to give improved damage estimates respect to the WL method (compare the two error indexes in Table 4.3).

On the contrary, better agreement with results from simulations seems to be associated with other spectral method. Namely, Dirlik, Zhao-Baker and the TB method show better agreement between the estimated rainflow damage and the damage computed from simulations, at least for slope k equal to 3 (see Figure 4.12).

For what concerns the Zhao-Baker technique, our calculations underlined that the simplest formulation, using Eqs. (4.14) and (4.15), is fairly inaccurate. In addition, for all spectra having $\alpha_2 = 0.1$ it incorrectly predicts a weight $w > 1$, which can not be true. On the contrary, the alternative improved version, Eq. (4.16) and (4.17) (that is presented in all figures) shows better results.

For the case of a S-N slope k equal to 5 a greater scatter amongst all results is observed (see Figure 4.13). In particular, it has to be underlined that for low values of the irregularity index α_2, the spread between the two bounds $\overline{D}_{\text{NB}}$ and $\overline{D}_{\text{RC}}$ of the rainflow damage defined in Eq. (4.42) is too high (ratio $\overline{D}_{RC}/\overline{D}_{NB}$ is about α_2^{k-1}), and this information is not sufficiently accurate for reliable fatigue damage assessment.

In conclusion, by taking the error index EI as a measure of the accuracy of damage estimations, all values reported in the last column of Table 4.3 clearly confirm that amongst all spectral methods, the most accurate ones are the ZB, DK, TB (both versions) and the empirical-$\alpha_{0.75}$ method, whereas the NB and WL techniques are markedly unreliable. In particular, the TB method has been shown to be as accurate as the DK approach, which is recognised as one of the best predictors for rainflow damage. However, the possibility of the TB method of estimating the $h_{RFC}(u, v)$ joint density gives the further advantage of including both the mean value influence and the effect of non-normalities (see Chapter 7).

In conclusion, we can say that the TB method is as accurate as the Dirlik method, which is recognised as the best predictor for rainflow damage. This does not hold for the WL method, which has shown great inaccuracy. Note that similar conclusions are confirmed by the systematic comparison of spectral methods carried out by [Gao and Moan 2008].

Chapter 5

FATIGUE ANALYSIS OF BIMODAL RANDOM LOADINGS

5.1. INTRODUCTION

Chapter 4 reviewed several methods for the fatigue analysis (i.e. estimation of the rain-flow cycle distribution and the fatigue damage) of broad-band random processes. It highlighted that the fatigue assessment procedure mainly relies on estimating the true distribution of counted cycles when rainflow method is used. In fact, in a broad-band random process, this distribution is not explicitly known, due to the complexity of the algorithm defining the rainflow count.

Amongst all Gaussian random processes, those having a spectral density concentrated around two well-separated frequencies represent a special class. Such processes are defined as bimodal processes and they can be schematised as the combination of two narrow-band processes. Due to this characteristic, we know in advance that cycles extracted by the rainflow method are approximately of two types, i.e. small and large cycles. Therefore, the estimation of the distribution of rainflow cycles becomes easier.

In the next sections, we shall review several spectral methods specifically developed for the fatigue analysis of bimodal Gaussian random processes: the single-moment (SM), the Jiao-Moan (JM), the Sakai-Okamura (SO) and the Fu-Cebon (FC) method. An original modification of the Fu-Cebon method (indicated as MFC method) will be proposed. Numerical simulations will be used to validate and compare all methods, included the Tovo-Benasciutti (TB) method developed for general broad-band processes.

We will see that the single-moment, the modified Fu-Cebon method and the TB method provide the best damage estimation.

5.2. SPECTRAL METHODS DEVELOPED FOR BIMODAL PROCESSES

Attention is focused on a special family $X(t)$ of zero-mean Gaussian processes that are the combination of two narrow-band processes with well-separated spectral components, see Figure 5.1.

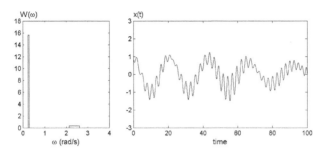

Figure 5.1: Sample of a bimodal Gaussian random process.

Such processes are referred to as bimodal processes and can be schematised as the sum of two independent narrow-band Gaussian processes:

$$X(t) = X_1(t) + X_2(t) \tag{5.1}$$

where $X_1(t)$ is the low-frequency (LF) and $X_2(t)$ the high-frequency (HF) component, associated to ω_1 and high ω_2 frequencies, respectively.

The spectral density function of $X(t)$ can be represented as the sum of two contributions:

$$W_X(\omega) = W_{X_1}(\omega) + W_{X_2}(\omega) \tag{5.2}$$

where $W_{X_1}(\omega)$ and $W_{X_2}(\omega)$ are the spectral densities of $X_1(t)$ and $X_2(t)$. The moments of the spectral density $W_X(\omega)$ are:

$$\lambda_m = \int_0^\infty \omega^m \, W_X(\omega) \, \mathrm{d}\omega = \lambda_{m,1} + \lambda_{m,2} \tag{5.3}$$

where $\lambda_{m,1}$ and $\lambda_{m,2}$ are the moments of $W_{X_1}(\omega)$ and $W_{X_2}(\omega)$.

The two processes $X_1(t)$ and $X_2(t)$ are narrow-band, with variances equal to $\lambda_{0,1}$ and $\lambda_{0,2}$, and with an intensity of mean upcrossings denoted as :

$$\nu_{0,1} = \frac{\omega_1}{2\pi} \qquad \nu_{0,2} = \frac{\omega_2}{2\pi} \tag{5.4}$$

The rainflow counting method applied to process $X(t)$ extracts two types of cycles: few large cycles (with amplitude s_L), associated to the interaction of $X_1(t)$ and $X_2(t)$ components, and many small cycles (with amplitude s_S) generated by the HF component $X_2(t)$ (see Figure 5.2).

77

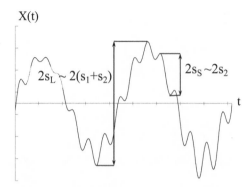

Figure 5.2: Cycles counted in a bimodal process by the rainflow method.

This observation is the basis of most approximate methods developed for estimating the amplitude distribution (denoted here as $p_{RFC}(s)$), counted in a bimodal process by the rainflow method. The differences amongst all methods are in the way the two contributions of both low and high components are taken into account.

As discussed in Chapter 3, if the $p_{RFC}(s)$ distribution is known, the fatigue damage rate under the linear rule (neglecting mean stress effects) is computed as:

$$\overline{D} = v_a \, C^{-1} \int_0^{+\infty} s^k \, p_{RFC}(s) \, ds \tag{5.5}$$

where v_a is the frequency of rainflow cycles and the S-N curve is $s^k N = C$.

5.2.1. Single-moment method (1990)

This method was formulated after examination of the trends observed in data from extensive simulation and rainflow analysis. The proposed formula for the rainflow damage intensity is:

$$\overline{D}_{RFC}^{SM} = \frac{2^{k/2}}{2\pi C} \, \Gamma\!\left(1 + \frac{k}{2}\right) \left(\lambda_{2/k}\right)^{k/2} \tag{5.6}$$

which depends on spectral moment $\lambda_{2/k}$. Formula above, proposed as an improvement of the narrow-band approximation, was shown to give accurate results in the case of bimodal spectra [Lutes and Larsen 1990, Larsen and Lutes 1991].

An interesting interpretation of SM spectral method in terms of multiaxial fatigue criteria has been recently given in [Cristofori et al. 2011b], by considering the sum of damage of uncorrelated loading projections in the five-dimensional deviatoric space.

78

5.2.2. Jiao-Moan method (1990)

Without loss of generality, let us first normalise process $X(t)$ so to define a new process $X^*(t)$ having a unit variance, i.e. [Jiao and Moan 1990]:

$$\lambda_0^* = \lambda_{0,1}^* + \lambda_{0,2}^* = 1 \tag{5.7}$$

where:

$$\lambda_1^* = \frac{\lambda_1}{\lambda_0} \qquad ; \qquad \lambda_2^* = \frac{\lambda_2}{\lambda_0} \tag{5.8}$$

are the variances of normalised $X_1^*(t)$ and $X_2^*(t)$ components, respectively. Process $X^*(t)$ have the same number of cycles as $X(t)$, except a suitable rescaling of their amplitudes.

The fatigue damage due to $X^*(t)$ is calculated as the sum of the damage given by large cycles, associated to the envelope $P(t)$ of process $X^*(t)$, and small cycles, associated to the envelope of the HF component $X_2^*(t)$.

The envelope $P(t)$ can be approximated as the sum of the normalised LF component $X_1^*(t)$ and the CL (Cramer-Leadbetter) envelope $R_2^*(t)$ of the normalised narrow-band HF component $X_2^*(t)$:

$$P(t) = X_1^*(t) + R_2^*(t) \tag{5.9}$$

In the Gaussian case, $X_1^*(t)$ and $R_2^*(t)$ follow a Gaussian and a Rayleigh distribution, respectively:

$$p_{X_1^*}(x) = \frac{1}{\sqrt{2\pi\,\lambda_{0,1}^*}} \exp\left(-\frac{x^2}{2\,\lambda_{0,1}^*}\right) \tag{5.10}$$

$$p_{R_2^*}(r) = \frac{r}{\lambda_{0,2}^*} \exp\left(-\frac{r^2}{2\,\lambda_{0,2}^*}\right) \tag{5.11}$$

The distribution of $P(t)$ is the convolution of distributions of $X_1^*(t)$ and $R_2^*(t)$:

$$p_P(x) = \int_{-\infty}^{+\infty} p_{X_1^*}(x-y)\,p_{R_2^*}(y)\,\mathrm{d}y$$

$$= \sqrt{\frac{\lambda_{0,1}^*}{2\pi}}\, e^{-\frac{x^2}{2\lambda_{0,1}^*}} + \sqrt{\lambda_{0,2}^*}\, x\, e^{-\frac{x^2}{2}}\, \Phi\left(\sqrt{\frac{\lambda_{0,2}^*}{\lambda_{0,1}^*}}\, x\right) \tag{5.12}$$

where $\Phi(\cdot)$ is the standard normal distribution function.

The time derivative $\dot{P}(t)$ is Gaussian and independent of $P(t)$, with zero mean and variance equal to:

$$\sigma_{\dot{P}}^2 = 4\pi^2 \left(\lambda_{0,1}^* v_{0,1}^2 + q_{X_2^*}^2 \lambda_{0,2}^* v_{0,2}^2 \right) \tag{5.13}$$

being $v_{0,1}$ and $v_{0,2}$ the mean upcrossing rates of processes $X_1^*(t)$ and $X_2^*(t)$, and $q_{X_2^*}$ the Vanmarcke's bandwidth parameter of $X_2^*(t)$. The upcrossing intensity spectrum of process $P(t)$ is approximate as:

$$v_P(x) = \frac{\sigma_{\dot{P}}}{\sqrt{2\pi}} p_P(x) \tag{5.14}$$

(i.e. the Rice's formula, valid in the limit of $P(t)$ and $\dot{P}(t)$ independent Gaussian processes). Then the expected mean upcrossing rate is:

$$v_{0,P} = v_P(x=0) = \lambda_{0,1}^* v_{0,1} \sqrt{1 + \frac{\lambda_{0,2}^*}{\lambda_{0,1}^*} \left(\frac{v_{0,2}}{v_{0,1}} q_{X_2^*} \right)^2} \tag{5.15}$$

The amplitude s_L of large cycles is calculated by referring to an amplitude process $Q(t)$ associated to $P(t)$, i.e.:

$$Q(t) = R_1^*(t) + R_2^*(t) \tag{5.16}$$

A plot of a sample of processes $X^*(t)$, $P(t)$ and $Q(t)$ is shown in Figure 5.3.

The distribution of $Q(t)$ is the convolution of distributions of $R_1^*(t)$ and $R_2^*(t)$ processes:

$$p_Q(q) = \int_{-\infty}^{+\infty} p_{R_1^*}(q-y)\, p_{R_2^*}(y)\, dy =$$

$$= \lambda_{0,1}^* q\, e^{-\frac{q^2}{2\lambda_{0,1}^*}} + \lambda_{0,2}^* q\, e^{-\frac{q^2}{2\lambda_{0,2}^*}} +$$

$$+ \sqrt{2\pi \lambda_{0,1}^* \lambda_{0,2}^*} \left(q^2 - 1 \right) e^{-\frac{q^2}{2}} \cdot \left[\Phi\left(\sqrt{\frac{\lambda_{0,1}^*}{\lambda_{0,2}^*}}\, q \right) + \Phi\left(\sqrt{\frac{\lambda_{0,2}^*}{\lambda_{0,1}^*}}\, q \right) - 1 \right] \tag{5.17}$$

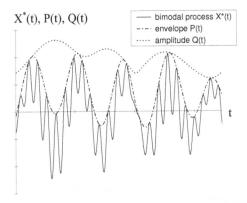

X*(t), P(t), Q(t)

——— bimodal process X*(t)
— · — envelope P(t)
· · · · · amplitude Q(t)

t

Figure 5.3: Samples of normalised bimodal process $X^*(t)$ (solid line), pseudo-envelope $P(t)$ (dashed line), envelope $Q(t)$ (dash-dotted- line).

Jiao and Moan propose to account for the contribution of $P(t)$ only when the LF component is important (i.e. only when $\lambda^*_{0,1}$ is large). Then, they further approximate Eq. (5.17):

$$p_Q(q) \cong \left(\lambda^*_{0,1} - \sqrt{\lambda^*_{0,1}\,\lambda^*_{0,2}}\right) q\, e^{-\frac{q^2}{2\lambda^*_{0,1}}} + \sqrt{2\pi\,\lambda^*_{0,1}\,\lambda^*_{0,2}}\left(q^2 - 1\right) e^{-\frac{q^2}{2}} \tag{5.18}$$

Note that Eq. (5.17) is an exact expression for $p_Q(x)$, while Eq. (5.18) is approximate and may give negative values when q approaches zero.

The large cycles associated to process $P(t)$ are assumed to have an amplitude $s_L = Q$, while their frequency of occurrence is approximated by $\nu_{0,P}$. Therefore, damage $\overline{D}_P$ of the large cycles generated by $P(t)$ is calculated as in Eq. (5.5) by substituting $p_Q(q)$ as the amplitude distribution $p_{S_L}(s)$, and $\nu_{0,P}$ as the frequency of occurrence. The use of Eq. (5.17) will provide an exact damage value $\overline{D}_P$, although not in closed-form, while the use of Eq. (5.18) will give the following approximate closed-form expression:

$$\overline{D}_P = \frac{\nu_{0,P}\,E\!\left[s_L^k\right]}{C}$$

$$= \frac{\nu_{0,P}}{C}\left(\sqrt{2}\right)^k \Gamma\!\left(1 + \frac{k}{2}\right)\left[\lambda_{0,1}^{*\left(\frac{k}{2}+2\right)}\left(1 - \sqrt{\frac{\lambda^*_{0,2}}{\lambda^*_{0,1}}}\right) + \sqrt{\pi\,\lambda^*_{0,1}\,\lambda^*_{0,2}}\,\frac{k\,\Gamma\!\left(\frac{k}{2}+\frac{1}{2}\right)}{\Gamma\!\left(1+\frac{k}{2}\right)}\right]$$

$$\tag{5.19}$$

The amplitude s_S of small cycles is taken as the amplitude associated to the HF component, which is Rayleigh distributed). The damage of small cycles is calculated by the
narrow-band approximation formula, i.e.:

$$\overline{D}_{X_2^*} = \frac{v_{0,2}}{C} \left(\sqrt{2\lambda_{0,2}^*} \right)^k \Gamma\left(1 + \frac{k}{2}\right) \tag{5.20}$$

The damage intensity for process $X^*(t)$ is then computed as the sum of previous two
contributions, namely the sum of Eqs. (5.19) and (5.20).

In order to simplify matter, this damage can be related to the damage computed according to the narrow-band approximation. More exactly, the damage intensity for normalised process $X^*(t)$ (which has $\lambda_0^* = 1$) computed by adopting the narrow-band approximation is:

$$\overline{D}_{X^*} = \frac{v_{0,X^*}}{C} \left(\sqrt{2} \right)^k \Gamma\left(1 + \frac{k}{2}\right) \tag{5.21}$$

being v_{0,X^*} the mean upcrossing intensity of $X^*(t)$, approximated as:

$$v_{0,X^*} \approx \sqrt{\lambda_{0,1}^* v_{0,1}^2 + \lambda_{0,2}^* v_{0,2}^2} \tag{5.22}$$

However, we know that for process $X^*(t)$ the damage intensity given by the narrow-
band approximation, Eq. (5.21), is a conservative estimate of the rainflow damage intensity. Thus, similarly to the scheme already adopted in the Wirsching-Light method
(see Chapter 4), a correction factor ρ_{JM} for the Jiao-Moan approach can be calculated
as the ratio of the approximate rainflow damage intensity (given by summing Eqs.
(5.24) and (5.25)) to the damage intensity obtained from the narrow-band approximation, Eq. (5.26):

$$\rho_{JM} = \frac{D_P + D_{X_2^*}}{D_{X^*}} =$$

$$= \frac{v_{0,P}}{v_0} \left[\lambda_{0,1}^* {}^{\left(\frac{k}{2}+2\right)} \left(1 - \sqrt{\frac{\lambda_{0,2}^*}{\lambda_{0,1}^*}}\right) + \sqrt{\pi \lambda_{0,1}^* \lambda_{0,2}^*} \frac{k\Gamma\left(\frac{k}{2}+\frac{1}{2}\right)}{\Gamma\left(1+\frac{k}{2}\right)} \right] + \frac{v_{0,2}}{v_0} \lambda_{0,2}^* \tag{5.23}$$

The rainflow damage intensity for the bimodal process is calculated as:

$$\overline{D}_{RFC}^{JM} = \rho_{JM} \overline{D}_{NB} \tag{5.24}$$

82

where $\overline{D}_{NB}$ is the narrow-band damage intensity and ρ_{JM} is defined in Eq. (5.23).

It is seen that ρ_{JM} is a function of the relative contributions from the two components ($\lambda_{0,1}^*$, $\lambda_{0,2}^*$, $\nu_{0,1}$, $\nu_{0,2}$), as well as the S-N curve slope k. Asymptotically $\rho_{JM} = 1$ when $\lambda_{0,1}^* = 1$ or $\lambda_{0,2}^* = 1$, which is the exact solution for a one-block narrow-band spectrum.

From simulation results, Jiao and Moan concluded that Eq. (5.23) is a good approximation for the ρ_{JM} correction factor (and then for the rainflow damage) for bimodal spectra having well-separated frequency contributions; in other situations, i.e. small $\nu_{0,2}/\nu_{0,1}$ values, the proposed analytical method may lead to $\rho_{JM} > 1$ for large $\lambda_{0,1}^*$ values. Such result derives from the fact that the simple sum of damage contributions from the pseudo-envelope process $P(t)$ and the faster process $X_2^*(t)$ may be too conservative; an alternative approach is then preferable.

5.2.3. Sakai-Okamura method (1995)

The method approximates the expected number $\overline{N}_L$ and $\overline{N}_S$ of large and small cycles counted in time T by the rainflow method by the number of mean upcrossings of $X_1(t)$ and $X_2(t)$ [Sakai and Okamura 1995]:

$$\overline{N}_L \cong \overline{N}_{0,1} = \frac{\omega_1}{2\pi}T \qquad ; \qquad \overline{N}_S \cong \overline{N}_{0,2} = \frac{\omega_2}{2\pi}T \qquad (5.25)$$

Therefore, the number of all cycles counted by the rainflow count is simply the sum of all cycles counted in each component, i.e.:

$$\overline{N}_{RFC} = \overline{N}_L + \overline{N}_S = T\frac{\omega_1 + \omega_2}{2\pi} \qquad (5.26)$$

In addition, the amplitude of large and small cycles, s_L and s_S, are assumed coincident with the amplitudes s_1 and s_2 associated to $X_1(t)$ and $X_2(t)$ narrow-band processes.

The consequence of this assumption is that the rainflow amplitude distribution is equal to a weighted linear combination of two amplitude distributions, corresponding to each component. Let $p_{RFC}^{SO}(s)$ be the amplitude density of rainflow cycles as predicted by Sakai-Okamura method. Given the amplitude s, let us express the expected number of rainflow cycles, say $\overline{N}(s)$, as the sum of cycles counted in each component, i.e.:

$$\overline{N}(s) = \overline{N}_{RFC}\, p_{RFC}^{SO}(s)\,\mathrm{d}s = \overline{N}_L\, p_{s_1}(s)\,\mathrm{d}s + \overline{N}_S\, p_{s_2}(s)\,\mathrm{d}s \qquad (5.27)$$

where $p_{s_1}(s)$ and $p_{s_2}(s)$ are the Rayleigh probability densities of the amplitudes s_1 and s_2 associated to the LF and HF narrow-band components. It follows that the rainflow amplitude distribution is:

83

$$p_{\text{RFC}}^{\text{SO}}(s) = \frac{\omega_1\, p_{s_1}(s)\, +\, \omega_2\, p_{s_2}(s)}{\omega_1 + \omega_2} \tag{5.28}$$

Fatigue damage is the sum of the damage caused by the low and high component:

$$\begin{aligned}
\overline{D}_{\text{RFC}}^{\text{SO}} &= \frac{1}{2\pi C}\left[\omega_1 \int_0^\infty s^k\, p_{s_1}(s)\,\mathrm{d}s + \omega_2 \int_0^\infty s^k\, p_{s_2}(s)\,\mathrm{d}s \right] \\
&= \frac{1}{2\pi C}\, \Gamma\!\left(1+\frac{k}{2}\right)\!\left[\omega_1 \left(\sqrt{2\lambda_{0,1}}\right)^k + \omega_2 \left(\sqrt{2\lambda_{0,2}}\right)^k \right]
\end{aligned} \tag{5.29}$$

and which is obviously equivalent to substitute $p_{\text{RFC}}^{\text{SO}}(s)$ density into Eq. (5.5), and using as intensity of counted cycles that derived from Eq. (5.26):

$$\nu_{\text{a}} = \frac{\omega_1 + \omega_2}{2\pi} \tag{5.30}$$

Finally, since the average frequency for each (narrow-band) component is:

$$\omega_1 = \left(\frac{\lambda_{2,1}}{\lambda_{0,1}}\right)^{1/2} \qquad , \qquad \omega_2 = \left(\frac{\lambda_{2,2}}{\lambda_{0,2}}\right)^{1/2} \tag{5.31}$$

the formula for the rainflow damage intensity according to Sakai and Okamura is [Sakai and Okamura 1995]:

$$\overline{D}_{\text{RFC}}^{\text{SO}} = \frac{2^{k/2}}{2\pi C}\, \Gamma\!\left(1+\frac{k}{2}\right)\!\left[\lambda_{0,1}^{(k-1)/2}\, \lambda_{2,1}^{1/2} + \lambda_{0,2}^{(k-1)/2}\, \lambda_{2,2}^{1/2}\right] \tag{5.32}$$

Note that previous damage formula reduces to the narrow-band approximation when either of the two frequency contributions vanishes.

5.2.4. Fu-Cebon method (2000)

The method assumes that the number $\overline{N}_{\text{L}}$ of large cycles equals the number of mean upcrossings of the LF component, i.e. $\overline{N}_{0,1} = \omega_1 T/2\pi$, and the number of the remaining small cycles, $\overline{N}_{\text{S}}$, is equal to $\overline{N}_{0,2} - \overline{N}_{0,1} = (\omega_2 - \omega_1)T/2\pi$, i.e. the larger cycles are not included in the amount of small cycles and are counted only once. Then, the expected number of rainflow cycles in time T is $\overline{N}_{\text{RFC}} = \overline{N}_{\text{L}} + \overline{N}_{\text{S}} = \overline{N}_{0,2}$, i.e. the number of mean upcrossings of the HF component.

Let $p_{s_1}(s)$ and $p_{s_2}(s)$ be the probability densities of amplitudes s_1 and s_2, respectively, and let $p_{\text{RFC}}^{\text{FC}}(s)$ be the distribution of rainflow amplitudes according to Fu-

84

Cebon method. Given amplitude s, let us express the number of rainflow cycles, say $\overline{N}(s)$, as the sum of cycles counted in each component, i.e.:

$$\overline{N}(s) = \overline{N}_{RFC}\ p_{RFC}^{FC}(s)\,ds = \overline{N}_L\ p_{s_L}(s)\,ds + \overline{N}_S\ p_{s_2}(s)\,ds \tag{5.33}$$

which gives the density of amplitudes of rainflow cycles as:

$$p_{RFC}^{FC}(s) = \frac{\overline{N}_{0,1}}{\overline{N}_{0,2}}\ p_{s_1}(s) + \left(1 - \frac{\overline{N}_{0,1}}{\overline{N}_{0,2}}\right) p_{s_2}(s) \tag{5.34}$$

Similarly to the JM method, the amplitude s_L of the large cycles is approximated as the sum of the amplitude of $X_1(t)$ and $X_2(t)$ components, i.e. $s_L \cong s_1 + s_2$, thus its distribution is given by the convolution of distributions $p_{S_1}(s)$ and $p_{S_2}(s)$ [Fu and Cebon 2000, Mood et al. 1987]:

$$
\begin{aligned}
p_{s_L}(s) &= \int_0^s p_{s_1}(s)\ p_{s_2}(s-y)\,dy \\
&= \frac{1}{\lambda_{0,1}\,\lambda_{0,2}}\,e^{-\frac{s^2}{2\lambda_{0,2}}} \int_0^s (s\,y - y^2)\,e^{-U\,y^2 + V\,s\,y}\,dy
\end{aligned}
\tag{5.35}
$$

where the following parameters are used:

$$U = \frac{1}{2\,\lambda_{0,1}} + \frac{1}{2\,\lambda_{0,2}} \quad , \qquad V = \frac{1}{\lambda_{0,2}} \tag{5.36}$$

Note that the use of a convolution corresponds to the hypothesis of independence between s_1 and s_2, which is valid only for well-separated frequency components.

The amplitude of small cycles is $s_S \cong s_2$, i.e. the amplitude associated to the HF narrow-band component $X_2(t)$, which has a Rayleigh density:

$$p_{s_S}(s) = \frac{s}{\lambda_{0,2}}\,e^{-\frac{s^2}{2\lambda_{0,2}}} \tag{5.37}$$

The total fatigue damage $D(T)$ is calculated by summing the damage contribution produced by both small and large cycles:

$$
\begin{aligned}
D_{RFC}^{FC}(T) &= \overline{N}_L \int_0^\infty \frac{s^k}{C}\,p_{s_L}(s)\,ds + (\overline{N}_S - \overline{N}_L) \int_0^\infty \frac{s^k}{C}\,p_{s_S}(s)\,ds \\
&= \frac{T}{2\pi\,C}\left[\omega_1 \int_0^\infty s^k\,p_{s_L}(s)\,ds + (\omega_2 - \omega_1) \int_0^\infty s^k\,p_{s_S}(s)\,ds \right]
\end{aligned}
\tag{5.38}
$$

Since the smallest cycles have amplitude s_2 which is Rayleigh distributed, their damage is the narrow-band approximation. Therefore, final form for the rainflow damage intensity according to Fu-Cebon method becomes [Fu and Cebon 2000]:

$$\overline{D}_{RFC}^{FC} = \frac{1}{2\pi C}\left[\frac{1}{\lambda_{0,1}\,\lambda_{0,2}} \int_0^\infty s^k e^{-\frac{s^2}{2\lambda_{0,2}}}\, I(s)\, ds \; + \right.$$

$$\left. + (\omega_2 - \omega_1)\left(\sqrt{2\lambda_{0,2}}\right)^k \Gamma\!\left(1+\frac{k}{2}\right)\right]$$

(5.39)

where $I(s)$ the integral:

$$I(s) = \int_0^s (s\,y - y^2)\, e^{-U y^2 + V s y}\mathrm{d}y$$

(5.40)

Formula above also follows by substituting the distribution given by Eq. (5.34) into Eq. (5.5), and by using $v_a = \overline{N}_{0,2}/T$ as the intensity of counted cycles.

5.2.5. Modified Fu-Cebon method [Benasciutti and Tovo 2007a]

All the methods presented in previous sections estimate the rainflow damage by adding the damage contributed by large and small cycles (associated to the low and high frequency component), the only difference being in the way each damage contribution is computed. In Table 5.1 we summarise the main quantities differentiating all such methods.

Both the JM and the FC methods account for the interaction between the LF and the HF component by identifying the amplitude s_L of large cycle as the sum of the amplitudes s_1 and s_2 of $X_1(t)$ and $X_2(t)$ processes (note that Sakai-Okamura method does not account for this interaction). The distribution $p_{s_L}(s)$, calculated as a convolution, is available either as an exact or an approximate formula, see Eqs. (5.17) and (5.18). Nevertheless, the Eq. (5.18) (i.e. the approximate formula) may give negative values; hence its exact version seems preferable.

The JM method associates the large cycles to the envelope process $P(t)$, thus it approximates their frequency $v_{a,L}$ by the mean upcrossing rate $v_{0,P}$ of process $P(t)$. This approximation seems quite correct, since it makes $v_{a,L}$ dependent on the relative contribution of the HF and LF components; for example, $v_{c,L}$ tends to decay to zero when $\lambda_{0,2}/\lambda_{0,1}$ increases. On the contrary, the FC method approximates the frequency of large cycles $v_{c,L}$ by $v_{0,1}$ (i.e. the mean upcrossing rate of $X_1(t)$), which does not depend on $\lambda_{0,2}/\lambda_{0,1}$ yet.

86

For what concerns small cycles, both methods approximate the amplitude S_S by the amplitude S_2 associated to the narrow-band HF component $X_2(t)$, and frequency $v_{a,S}$ by $v_{0,2}$, i.e. the mean upcrossing rate of $X_2(t)$.

Table 5.1: Comparison of spectral method specific for bimodal processes.

Spectral method	Large cycles		Small cycles	
	Cycle frequency $v_{a,L}$	Amplitude density	Cycle frequency $v_{a,S}$	Amplitude density
JM	$v_{0,P}$	$p_Q(s)$ approximate, Eq. (5.18)	$v_{0,2}$	$p_{s_2}(s)$
SO	$v_{0,1}$	$p_{s_1}(s)$	$v_{0,2}$	$p_{s_2}(s)$
FC	$v_{0,1}$	$p_Q(s)$ exact, Eq. (5.17)	$v_{0,2} - v_{0,1}$	$p_{s_2}(s)$
MFC	$v_{0,P}$	$p_Q(s)$ exact, Eq. (5.17)	$v_{0,2} - v_{0,P}$	$p_{s_2}(s)$

However, not all cycles associated to $X_2(t)$ have small amplitudes, since some of them have peaks belonging to large cycles' amplitudes (see Figure 5.2). Therefore, as done by the FC method, such large cycles must be subtracted from the total number of small cycles. Updating this argument to $v_{0,P}$, assumed as the correct frequency of large cycles, provides $v_{0,2} - v_{0,P}$ as the actual frequency small cycles.

In conclusion, we suggest to slightly modify the FC method by using $v_{0,P}$ as the frequency of large cycles (whose amplitude s_L is distributed according to $p_Q(s)$, which can be computed by the exact formula reported is in Eq. (5.17)), and at the same time to adopt $v_{0,2} - v_{0,P}$ for the frequency of small cycles.

$$\overline{D}_{\text{RFC}}^{\text{MFC}} = v_{0,P} \int_0^\infty \frac{s^k}{C} p_Q(s) \, ds + (v_{0,2} - v_{0,P}) \int_0^\infty \frac{s^k}{C} p_{s_2}(s) \, ds \qquad (5.41)$$

which requires numerical integration of the first integral contribution. Numerical simulations will show how Eq. (5.41) is more accurate than the FC method.

5.3. NUMERICAL SIMULATIONS

Numerical simulations are used to generate various Gaussian random processes with a bimodal spectral density; two types of spectral densities are investigated: ideal bimodal spectra and a more realistic spectrum.

Ideal bimodal spectral densities, also used in [Lutes and Larsen 1990] are built by adding together two rectangular blocks (see Figure 5.4). This type of spectra allows us to relate the accuracy of the damage estimation to some geometrical parameters defining the shape of the bimodal spectrum, as well as to α_1 and α_2 indexes.

In the second part, a more realistic bimodal spectral density is used, as that given in [Larsen and Lutes 1990] and used by Wirsching (see [Wirsching and Light 1980]).

As usual, for each bimodal spectral density, a sufficiently long time history is simulated in time domain; cycles are then extracted by the rainflow method and fatigue damage is calculated under the Palmgren-Miner linear damage rule.

The damage intensity from simulated time histories is compared with predictions made in the frequency-domain by the various spectral methods previously reviewed.

Damage comparison will assume for the S-N curve a reference fatigue strength $C = 1$ and a slope equal to $k = 3$ and $k = 5$, respectively.

5.3.1. Ideal bimodal spectral densities

Let us firstly concentrate to the family of ideal bimodal spectral densities, constructed as superposition of two rectangular blocks, see Figure 5.4. These spectra are characterised by the amplitude levels W_1 and W_2, and by the frequency ranges $\omega_a - \omega_b$ and $\omega_c - \omega_d$, which are related through the frequency ratio:

$$R = \frac{\omega_c}{\omega_a} = \frac{\omega_d}{\omega_b} \tag{5.42}$$

and the area ratio:

$$B = \frac{A_2}{A_1} = \frac{W_2 (\omega_d - \omega_c)}{W_1 (\omega_b - \omega_a)} = \frac{R W_2}{W_1} \tag{5.43}$$

in which A_1 and A_2 are the areas of the low and high block, respectively. We shall assume in all simulations $A_1 + A_2 = 1$ (i.e. variance is equal to unity).

These ideal bimodal spectral densities are particularly convenient for quantifying the relative contribution on fatigue damage of LF and HF components. In fact, if one neglects either of the two blocks, the spectral density becomes a single rectangular block. Further, if we set the frequency ratio for each rectangular block equal to:

$$c = \frac{\omega_b}{\omega_a} = \frac{\omega_d}{\omega_c} = \frac{1.1}{0.9} \tag{5.44}$$

then each of the single-block approaches a narrow-band spectrum.

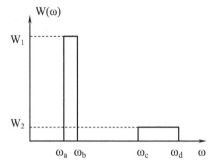

Figure 5.4: One-sided bimodal spectral density used for numerical simulations.

Other quantities defining these bimodal spectra are the central frequencies:

$$\omega_1 = \frac{\omega_a + \omega_b}{2} \quad , \quad \omega_2 = \frac{\omega_c + \omega_d}{2} \tag{5.45}$$

Once we choose the lowest frequency ω_a and the values of R and B, the levels W_1 and W_2 immediately follow, so that a unique spectral density is defined.

Spectral moments for this type of bimodal spectra are expressed as:

$$\lambda_m = \lambda_{m1} \left(1 + B R^m \right) \tag{5.46}$$

where λ_{m1} denotes the spectral moments for the low-frequency block. The formula for computing $\lambda_{2/k}$ moment, used in Eq. (5.6), is:

$$\lambda_{2/k} = \frac{k}{k+2} \frac{\omega_a^{2/k}}{(c-1)(1+B)} \left(c^{(k+2)/k} - 1 \right) \left(1 + B R^{2/k} \right) \tag{5.47}$$

where ω_a is the lowest frequency and c is the one-block frequency ratio.

It is possible to express α_1 and α_2 bandwidth parameters for process $X(t)$ as:

$$\alpha_1 = \sqrt{\frac{3 \left(c^2 - 1 \right)^2 \left(1 + B R \right)^2}{4 \left(c - 1 \right) \left(c^3 - 1 \right) \left(1 + B \right) \left(1 + B R^2 \right)}}$$

$$\alpha_2 = \sqrt{\frac{5 \left(c^3 - 1 \right)^2 \left(1 + B R^2 \right)^2}{9 \left(c - 1 \right) \left(c^5 - 1 \right) \left(1 + B \right) \left(1 + B R^4 \right)}}$$

$$\tag{5.48}$$

and analogously β_1 and β_2 bandwidth parameters relative to the derivative process $\dot{X}(t)$ as:

89

$$\beta_1 = \sqrt{\frac{15\,(c^4-1)^2\,(1+B\,R^3)^2}{16\,(c^3-1)\,(c^5-1)\,(1+B\,R^2)\,(1+B\,R^4)}}$$

(5.49)

$$\beta_2 = \sqrt{\frac{21\,(c^5-1)^2\,(1+B\,R^4)^2}{25\,(c^3-1)\,(c^7-1)\,(1+B\,R^2)\,(1+B\,R^6)}}$$

Note that, since c is a constant, the above expressions are functions only of R and B parameters; therefore, for a given c value (as that in Eq. (5.44)), an appropriate choice of R and B uniquely defines α_1 and α_2 values, the other β_1 and β_2 following as a consequence.

The set of simulations using ideal bimodal spectral densities is further subdivided in two parts: attention is firstly focused on R and B parameters, and secondly on α_1 and α_2 bandwidth parameters. Since most methods specifically developed for bimodal processes are valid only under certain hypotheses (e.g. when the two components are sufficiently far apart), the first set of simulations aims to find the optimal ranges of R and B indexes for each method.

5.3.1.1. Simulations depending on R and B parameters

The R parameter (i.e. the frequency ratio) varied from 1 (i.e. two components completely overlapped) up to 22 (i.e. well-separated components), while the area ratio B varied from 10^{-4} (only the lower narrow-band component) up to 10^3 (only the upper narrow-band component), see Table 5.2. The lowest frequency, arbitrarily set to $\omega_a = 24$ rad/sec, was verified to only affect the magnitude and not the relative values of damage results.

Table 5.2: Geometrical parameters of ideal bimodal spectra used in simulations.

R	B	ω_a [rad/sec]
$1 - 22$	$10^{-4} - 10^3$	24

The correlation between α_1 and α_2 indexes and R and B parameters is shown in Figure 5.5. When bimodal spectra reduce to a single rectangular block (narrow-band) spectrum (i.e. very low or very high B values), α_1 approaches unity very rapidly, whereas α_2 does not (especially for low values of B), which seems a quite surprisingly fact. However, a bimodal process having a spectrum with a quite small energy contribution at high frequencies (given for example by simultaneous low B and high R values) generates time histories having small oscillations superimposed on a slowly-varying component, having a low irregularity index α_2.

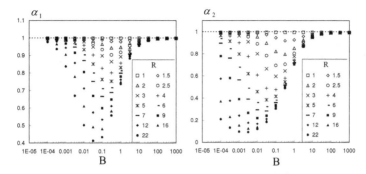

Figure 5.5: Correlation existing between α_1 and α_2 and R and B geometrical parameters.

Results presented in the next figures are given as the ratio $\overline{D}_{RFC}^*/D_{RFC}$ of the theoretical to the numerically computed damage, i.e. they are dimensionless. Damage $\overline{D}_{RFC}^*$ is the value computed by any of the spectral methods previously reviewed.

Results relative to a S-N slope $k = 3$ are summarised from Figure 5.6 through Figure 5.16. Results relative to the slope $k = 5$ are not reported, since they are quite similar to those observed for $k = 3$, even if all deviations between damage from simulations and estimated damage are more pronounced.

The narrow-band approximation is quite good, as expected, in rather narrow-band cases, that is when B is very high or very low (independently of R), or alternatively when R approaches unity (independently of B). In all other situations (i.e. when spectra become more broad-banded), it gives predictions markedly worse, as in cases with large R values and intermediate B values (see Figure 5.6) For example, a deviation of as much as 218 percent for $k = 3$ is observed when $R = 22$ and $B = 0.1$, corresponding to the most broad-band spectrum, with $\alpha_1 = 0.433$ and $\alpha_2 = 0.306$).

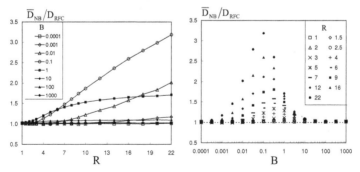

Figure 5.6: Damage estimation by the narrow-band approximation. Ideal bimodal spectral density. S-N slope: $k = 3$.

Compared to the narrow-band approximation, the SM method provides in general more accurate estimates, see Figure 5.7. For $k=3$, the largest deviations occur for R lying in the range $3-7$ and B in the range $0.1-1$, being of about negative 12 percent ($R=5$ and $B=0.1$). When the bimodal spectra become narrow-band, the SM method converges to the narrow-band approximation.

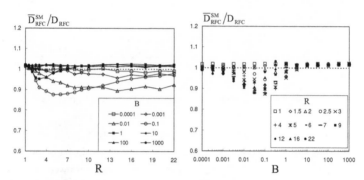

Figure 5.7: Damage estimation by the SM method. Ideal bimodal spectral density. S-N slope: $k=3$.

Results from the JM are presented in Figure 5.8 and Figure 5.9. In some cases this method gives $\rho_{JM} >1$, corresponding to $\overline{D}_{RFC}^{JM} > \overline{D}_{NB}$, which cannot be true (see Chapter 3). This situation generally happens for small R values ($R<4$) and for B in the range $0.01-1$. Results concerning damage predictions are reported in Figure 5.9 only when the condition $\rho_{JM} \le 1$ is met.

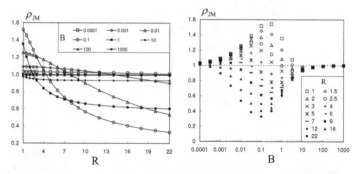

Figure 5.8: Damage correction factor ρ_{JM} defined by the JM method. Ideal bimodal spectral density. S-N slope: $k=3$.

As can be seen, the JM method is consistent with the narrow-band approximation (i.e. $\rho_{JM} =1$) for very large or very low B values. In all other cases, it generally slightly overestimates rainflow damage, in particular when $B=0.01-1$ (independently of R). The maximum positive deviation, of bout 26 percent, occurs for $R = 1.5$ and $B = 0.01$.

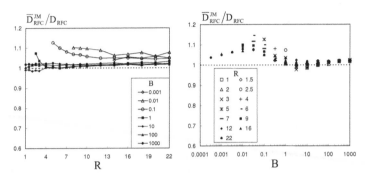

Figure 5.9: Damage prediction by the JM method. Ideal bimodal spectral density. S-N slope: $k = 3$.

Results from the SO method are sketched in Figure 5.10. The method is expected to provide good predictions either when the bimodal spectrum reduces to a narrow-band spectrum or for well-separated frequencies ω_1 and ω_2 (i.e. for high R values). The first condition is always satisfied (i.e. very low or very high B values), while the second one is never met, since no asymptotic behaviour against R is revealed.

What is quite surprising is, instead, a marked dependence of damage estimation on the area ratio B, independently of R. In particular, predictions become worse (i.e. rain-flow damage is always underestimated) for B in $0.01 - 10$ interval, with a maximum difference of about negative 36 percent when $R = 4$ and $B = 0.3$. Note how this result is also present in the original paper by Sakai and Okamura (e.g. see Figure 10 in [Sakai and Okamura 1995]).

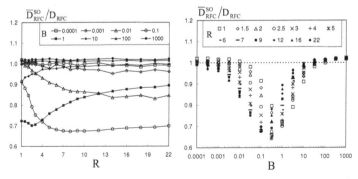

Figure 5.10: Damage estimation by the SO method. Ideal bimodal spectral density. S-N slope: $k = 3$.

This inaccuracy is probably related to an incorrect estimation of amplitude distribution $p_{s_L}(s)$ of large cycles, which gives a damage contribution lower than observed. In fact the SO method approximates $p_{s_L}(s)$ by the Rayleigh probability density $p_{s_2}(s)$, valid for the amplitude of cycles associated to $X_2(t)$, and thus completely neglects the contribution of superimposed small amplitudes (as done by FC and JM methods), resulting from the interaction between $X_1(t)$ and $X_2(t)$.

As an example, Figure 5.11 shows how the SO method attributes less probability to higher amplitudes than done by the FC and JM methods. The same figure shows how the approximate density in the JM method, i.e. Eq. (5.18), might give negative values for small amplitudes), the difference increasing with B (i.e. when the contribution of small amplitudes makes this difference more evident).

At the opposite, the FC method shows a damage prediction which is highly dependent on the frequency ratio R, see Figure 5.12. In particular, the accuracy considerably improves for well-separated frequency components, e.g. R values greater than 7 at least (in fact this assures that hypothesis of independence between s_1 and s_2 amplitudes used in deriving $p_{s_L}(s)$ density, as well as the rainflow damage, is asymptotically satisfied).

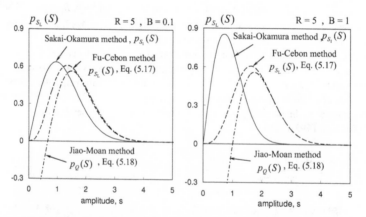

Figure 5.11: Amplitude distribution $p_{s_L}(s)$ of large cycles. Comparison is amongst SO, FC and JM method. Note negative values taken by the approximate density used by the JM method.

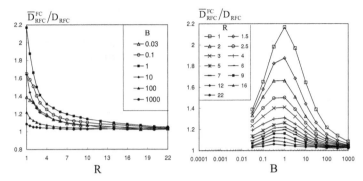

Figure 5.12: Damage estimation by the FC method. Ideal bimodal spectral density. S-N slope: $k = 3$.

As already noted, the method requires numerical integration, which cannot be achieved for too low $\lambda_{0,2}$ values, as given by low B values (e.g. integration may be impossible when $B < 0.03$, for all values of R, see Figure 5.12).

Figure 5.12 confirms that the FC method correctly converges to the narrow-band approximation when the bimodal spectra reduce to a single rectangular block (low and high B values). On the contrary, in all other cases the predictions generally overestimate the rainflow damage computed in simulations and they are quite good only for sufficiently high R values, e.g. the largest deviation (for $k = 3$) is about 41 percent for $R \geq 3$, about 22 percent for $R \geq 6$ and it is always below 15 percent when $R \geq 9$.

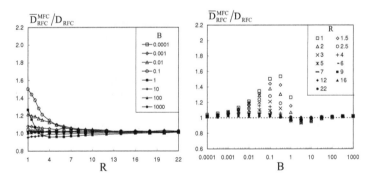

Figure 5.13: Damage estimation by the MFC method. Ideal bimodal spectral density. S-N slope: $k = 3$.

The variant introduced by the MFC method seems to considerably improve the ability of damage prediction of FC method, see Figure 5.13. First, the damage calculation, making use of the exact formula given in Eq. (5.17) for $p_{s_L}(s)$ density, does not present problems in numerical integration. Further, the MFC technique provides more accurate damage estimations, since it converges more rapidly to high accuracies when R increases, with deviations (for $k = 3$) below 21 percent for $R \geq 3$, below 9 percent for $R \geq 6$ and below 5 percent when $R \geq 9$.

Comparisons regarding the TB method are shown in Figure 5.14: good agreement is generally observed for all the values taken by R and B parameters, being the largest deviation about negative 6 percent.

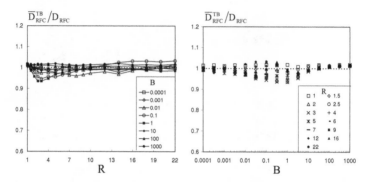

Figure 5.14: Damage prediction by the TB method. Ideal bimodal spectral density. S-N slope: $k = 3$.

If compared to results provided in previous figures, the TB method seems much more accurate than the other spectral methods.

A possible explanation of this high accuracy may be as follows. As said in Chapter 4, the TB method estimates the rainflow damage making use of an approximate expression for the range count damage (i.e. the lower bound of the rainflow damage).

The approximate formula for range count damage is asymptotically correct only when amplitudes and mean values of range counted cycles are independent (with a Rayleigh and Gaussian distribution, respectively). This condition should be actually satisfied in bimodal processes having well-separated components: range counted cycles in fact correspond to all small cycles generated by the HF component, which have Rayleigh distributed amplitudes and whose mean values are coincident with the values of the LF component, which is Gaussian.

In order to prove this assertion, the ratio of the range count damage intensity $\overline{D}_{RC}$, as given by Eq. (4.34) to that computed from simulated time histories has been investigated. As already seen in Chapter 4, Figure 5.16 shows that the accuracy of the approximate formula for the range count damage virtually depends on the bandwidth difference $\Delta = \alpha_1 - \alpha_2$, being markedly worse when α_1 and α_2 significantly differ.

This result, unexpected for bimodal processes, shows that the cycle distribution of range count cycles should not neglect an explicit dependence on α_1 parameter.

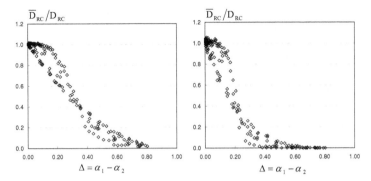

Figure 5.15: Comparison of the range count damage intensities: the approximate formula $\overline{D}_{RC}$ and the values D_{RC} from simulated time histories. S-N slope: $k = 3$ (left) and $k = 5$ (right).

On the other hand, the range of $\Delta = \alpha_1 - \alpha_2$ values in which the approximation does not work corresponds to b_{app} (i.e. the weighting factor entering Eq. (4.48)) converging to unity. This means that when the formula for the lower bound is fairly inaccurate, its contribution to the rainflow damage progressively diminishes.

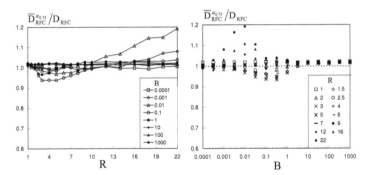

Figure 5.16: Damage estimation by the empirical-$\alpha_{0.75}$ method. Ideal bimodal spectral density. S-N slope: $k = 3$.

The rainflow damage as predicted by the empirical-$\alpha_{0.75}$ method is shown in Figure 5.16, and it is in quite close agreement with simulations.

5.3.1.2. Simulations depending on α_1 and α_2

The previous set of simulations considered bimodal spectral densities with different combinations of R and B indexes, while now we present other simulations in with there are ideal bimodal spectra having prescribed values of α_1 and α_2 parameters.

More exactly, for each pair of α_1 and α_2 indexes, a unique solution of the non-linear system in Eq. (5.48) is obtained in term of B and R variables (since c is constant). The complete set of such geometric parameters is given in Table 5.3, in conjunction with the corresponding values of α_1 and α_2 parameters.

Table 5.4 gives instead the values of β_1 and β_2 indexes for the derivative process. An empty value is given where the non-linear system has no solution. The variance of processes $X(t)$ has been set equal to unity.

Table 5.3: Geometrical parameters R and B defining ideal bimodal spectral densities used in numerical simulations.

α_2	α_1 values															
	0.450		0.500		0.550		0.600		0.700		0.800		0.850		0.900	
	R	B	R	B	R	B	R	B	R	B	R	B	R	B	R	B
0.800															3.101	1.265
0.750															2.591	0.627
0.700															2.539	0.318
0.650															2.696	0.171
0.600													3.233	0.240	2.981	0.098
0.550											3.978	0.245	3.412	0.147	3.366	0.059
0.500											4.093	0.158	3.728	0.092	3.849	0.037
0.450											4.387	0.101	4.176	0.058	4.447	0.024
0.400									6.063	0.123	4.855	0.065	4.774	0.037	5.194	0.015
0.350									6.454	0.080	5.527	0.041	5.565	0.023	6.148	0.010
0.300							9.248	0.073	7.176	0.050	6.470	0.025	6.632	0.014	7.408	0.006
0.250					11.715	0.051	10.140	0.045	8.338	0.030	7.827	0.015	8.132	0.009	9.154	0.004
0.200					13.292	0.029	11.855	0.025	10.206	0.017	9.888	0.008	10.379	0.005	11.751	0.002
0.150					16.466	0.015	15.037	0.013	13.429	0.008	13.340	0.004	14.113	0.002	16.046	0.001
0.140	21.281	0.019	19.442	0.014												
0.130																
0.120	23.450	0.013	21.961	0.010												
0.110																
0.100	26.585	0.009	25.593	0.007	23.346	0.006	21.730	0.005	19.988	0.003					24.586	0.0004

Table 5.4: Values of β_1 and β_2 bandwidth parameters corresponding to α_1 and α_2 given in Errore. L'origine riferimento non è stata

α_2	α_1 values															
	0.900		0.850		0.800		0.700		0.600		0.550		0.500		0.450	
	β_1	β_2	β_1	β_2	β_1	β_2	β_1	β_2	β_1	β_2	β_1	β_2	β_1	β_2	β_1	β_2
0.800	0.981	0.963														
0.750	0.963	0.851														
0.700	0.941	0.803														
0.650	0.916	0.752	0.931	0.865												
0.600	0.888	0.701	0.907	0.818												
0.550	0.858	0.653	0.880	0.770	0.940	0.830										
0.500	0.826	0.610	0.852	0.729	0.919	0.798										
0.450	0.793	0.619	0.822	0.691	0.895	0.824										
0.400	0.758	0.580	0.790	0.656	0.869	0.784	0.933	0.888								
0.350	0.721	0.546	0.756	0.625	0.841	0.746	0.913	0.855								
0.300	0.684	0.516	0.721	0.597	0.811	0.709	0.890	0.822	0.942	0.852						
0.250	0.645	0.489	0.684	0.569	0.779	0.677	0.865	0.789	0.923	0.832	0.945	0.864				
0.200	0.604	0.466	0.646	0.544	0.746	0.650	0.838	0.758	0.902	0.851	0.927	0.843				
0.150	0.563	0.446			0.711	0.622	0.809	0.731	0.879	0.822	0.907	0.822	0.926	0.846		
0.140													0.918	0.837		
0.130													0.910	0.829		
0.120															0.949	0.868
0.110															0.942	0.861
0.100	0.520	0.429					0.799	0.708	0.854	0.795	0.884	0.802	0.910	0.829	0.936	0.855

As stated by Eqs. (5.48) and (5.49), there is a strict dependence between the four bandwidth parameters α_1, α_2, β_1 and β_2 and the geometrical parameters R and B. This dependence can be best understood by the graphical representation shown in Figure 5.17 and Figure 5.18. Consequently, we may expect that these new simulations will give results similar to those already presented.

In fact, also these simulations have generally confirmed the trends observed in previous results. Anyway, some results are particularly interesting and need some comments.

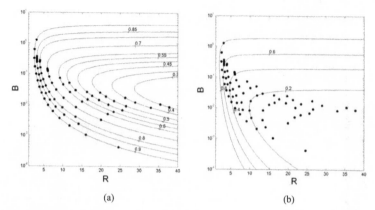

Figure 5.17: Relationship between B and R indexes and (a) α_1, (b) α_2 bandwidth parameters for ideal bimodal spectral density. Dots represent all pairs used in numerical simulations.

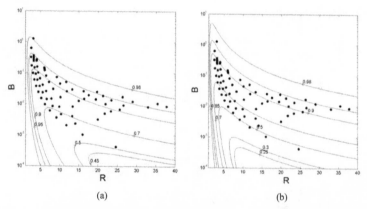

Figure 5.18: Relationship between B and R indexes and (a) β_1, (b) α_2 bandwidth parameters for ideal bimodal spectral density. Dots represent all pairs used in numerical simulations.

In particular, Figure 5.19 shows how the narrow-band approximation is more sensitive to α_1 than α_2; in fact, the prediction becomes quite accurate only when α_1 converges to unity, independently of α_2.

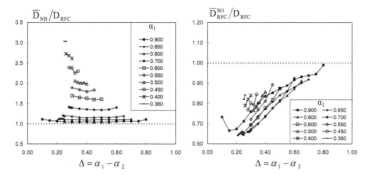

Figure 5.19: Damage estimation by the narrow-band approximation (left) and the SO method (right). Ideal bimodal spectral density. S-N slope: $k = 3$.

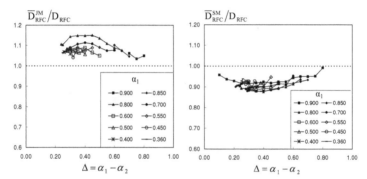

Figure 5.20: Damage estimation by the JM method (left) and the SM method (right). Ideal bimodal spectral density. S-N slope: $k = 3$.

In addition, Figure 5.21 confirms use how the MFC method provides in general more accurate damage estimations than the FC method.

Finally, Figure 5.22 shows that the TB method is quite accurate as in previous simulations.

101

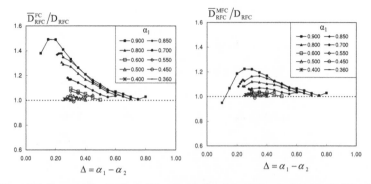

Figure 5.21: Damage estimation by the FC method (left) and the MFC method (right). Ideal bimodal spectral density. S-N slope: $k = 3$.

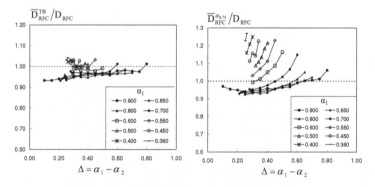

Figure 5.22: Damage estimation by the TB method (left) and the empirical-$\alpha_{0.75}$ method (right). Ideal bimodal spectral density. S-N slope: $k = 3$.

5.3.2. More realistic bimodal spectra

The ideal bimodal spectral densities used in previous simulations gave the possibility to get important information about the efficiency of several spectral methods, although they cannot represent actual spectra encountered in practice.

In subsequent simulations, a more realistic bimodal spectral density is considered, expressed as:

$$W_X(\omega) = A \frac{7793}{T_D \, \omega^5} e^{-\frac{1948}{T_D^4 \, \omega^4}} \cdot \frac{1}{\left[1 - \left(\dfrac{\omega}{\omega_N}\right)^2\right]^2 + 4\xi\left(\dfrac{\omega}{\omega_N}\right)^2} \tag{5.50}$$

where A is a scale factor, T_D is the dominant wave period and T_N is the dominant structural period, and ξ is the structural damping coefficient. If we set $\tau = \omega_D/\omega_N$, being $\omega_N = 2\pi/T_N$ and $\omega_D = 2\pi/T_D$ (with $\omega_N \geq \omega_D$), we can describe these spectra by the ratio $R = 1/\tau$ of the high to the low frequency; the area ratio $B = \lambda_{0,2}/\lambda_{0,1}$ of the high to the low area is governed by the damping coefficient ξ. The parameter values are reported in Table 5.5: the τ ratio has been varied in the range $0.045 - 0.500$, corresponding to a R value lying in the range $2 - 22$ (note that two-peaked spectra are obtained only if τ does not exceed 0.500); the damping coefficient ξ varied between 0.005 and 0.080. T_D is equal to 5 sec for all spectra. The A factor is chosen so to always have spectra with a unit area.

Table 5.5: Geometrical parameters of more realistic spectra used in simulations.

T_D [sec]	$\tau = \omega_D/\omega_N$	ξ
5	0.045-0.500	0.005-0.080

As shown in Figure 5.23, the ratio $B = \lambda_{0,2}/\lambda_{0,1}$ of the high to the low area is governed by the damping ratio ξ. The correlation between ξ and B is shown in Figure 5.24.

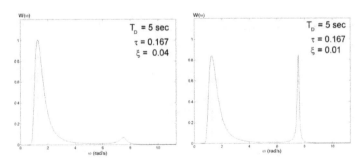

Figure 5.23: More realistic one-sided bimodal spectral density used for numerical simulations.

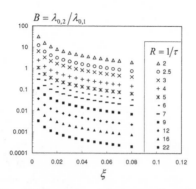

Figure 5.24: Correlation existing between ξ and B.

All simulations relative to this type of bimodal spectral densities substantially confirm previous results, obtained with ideal bimodal spectra.

For example, the narrow-band approximation and the SO method are in general not quite accurate, see Figure 5.25, being very dependent on the spectrum parameters τ and ξ.

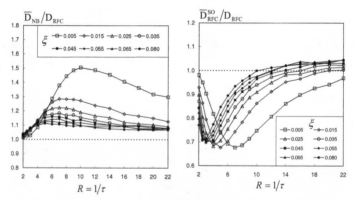

Figure 5.25: Damage estimation by the narrow-band approximation (left) and the SO method (right). More realistic bimodal spectra. S-N slope: $k = 3$.

On the contrary, the SM method shows predictions in good agreement with simulations (see Figure 5.26), with the largest deviation is about negative 15 percent, while for the TB method the maximum deviation is about negative 7 percent.

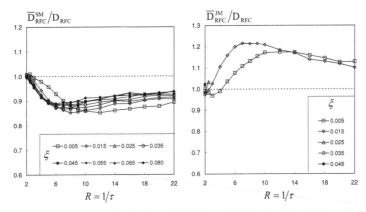

Figure 5.26: Damage estimation by the SM method (left) and the JM method (right). More realistic bimodal spectra. S-N slope: $k = 3$.

For what concerns the JM method (see Figure 5.26), for most spectra it is $\rho_{JM} > 1$, and when $\rho_{JM} \leq 1$ predictions are generally greater than results from simulations (the largest deviation of about positive 24 percent).

Results from the FC method and its modification are shown in Figure 5.27, where it is shown how the FC method generally overestimates the rainflow damage from simulations, being accurate only for sufficiently high R values. Furthermore, the modification of the original FC method improves considerably the prediction: the largest deviation for the original FC method (for $k = 3$) is about 68 percent for $R \geq 3$, about 37 percent for $R \geq 6$ and below 24 percent when $R \geq 9$, whereas the modified FC technique gives a deviation of about 32 percent for $R \geq 3$, about 25 percent for $R \geq 6$ and below 17 percent when $R \geq 9$.

In Figure 5.28 we show the results of the method based on $\alpha_{0.75}$ parameter (the largest error is about negative 8 percent).

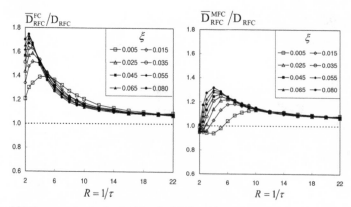

Figure 5.27: Damage estimation by the FC method (left) and the MFC method (right). More realistic bi-modal spectra. S-N slope: $k = 3$.

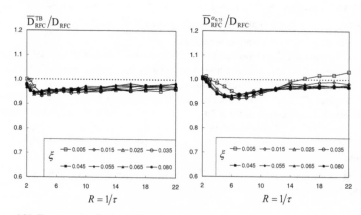

Figure 5.28: Damage estimation by the TB method (left) and the empirical-$\alpha_{0.75}$ method (right). More realistic bimodal spectra. S-N slope: $k = 3$.

Chapter 6

A SIMPLE CASE STUDY: A CAR-QUARTER MODEL MOVING ON AN IRREGULAR ROAD

6.1. INTRODUCTION

One of the main advantages of spectral methods is the possibility to use a frequency-domain approach for fatigue life assessment of real mechanical components subjected to external random loadings. Spectral methods are used in conjunction with the traditional frequency-domain response analysis, that can be performed for example with the Finite Element Method or the Multi-body analysis, see [Haiba et al. 2002].

A typical situation is represented by automotive applications concerning the simulation of the dynamical behaviour of two- or four-wheeled vehicles subjected to road irregularity. The road is modelled as a random process with known spectral content and in the hypothesis of a linear analysis, the frequency content of all internal responses can be calculated by a classical frequency-domain analysis.

From the knowledge of the response spectral density, on can easily use consolidated spectral methods to assess, for example, the influence on the estimated damage and related fatigue life determined by a change of one model parameter. This approach, obviously, seems very attractive in predicting the fatigue life of mechanical components at the design stage.

In order to easily derive the internal response spectrum from the external excitation spectrum, the frequency-domain analysis requires a linear model, then excluding non-linear springs and dampers models, i.e. $k = k(x)$ and $c = c(y)$, where x and y are some variables determining the current state of the system. Furthermore, another strong hypothesis is that the vehicle is assumed constantly in contact with the road, thus the car-quarter movement over holes has to be treated differently (see for example [Bogsjö 2002]).

This Chapter illustrates a simple case study, i.e. a discrete-parameter model moving on a irregular road, which is modelled as a spatial Gaussian random process $\zeta(x)$, with known spectral density. For this simple model, the frequency-domain dynamic analysis provides explicit analytical formulas for the response spectral density, under the hypothesis of linearity. If we require a more accurate description of the dynamic behaviour of the system, more complex dynamic models for the car can be adopted, which however increase the computational effort (see [Haiba et al. 2002]).

6.2. ROAD ROUGHNESS DESCRIPTION

Road surfaces can be represented as realisations of random processes, provided that the effect of such occasional large irregularities as potholes are removed from the analysis and treated separately. The simplest approach assumes the road irregularity as a realisation of a two-dimensional Gaussian random process, which, in the hypothesis of being homogenous and isotropic, is uniquely characterised by its spectral density [Dodds and Robson 1973, Robson and Dodds 1975/76].

Let us model the height of the road profile as a realisation of a spatial stationary Gaussian random process $\zeta(x)$. The height $\zeta(x)$ does not depend on time variable t, while it is a function of the spatial distance x along the road. Low frequency components correspond to long wavelength roughness and high frequency components correspond to short wavelength roughness.

In the hypothesis the process is homogenous and isotropic, a one-sided spectral density, say $W_\zeta(n)$ [m^3/cycle], uniquely characterises the frequency content of process $\zeta(x)$. The quantity $n = 1/\lambda$ [cycles/m] is the wave number, being λ the road wavelength.

When a vehicle travels along the road at a constant speed v, the uniform Gaussian random process $\zeta(x)$ in spatial domain is converted into a stationary Gaussian random process in time domain, say $z(t)$, which excites the vehicle. The one-sided (angular frequency) spectrum $W_z(\omega)$ [(m^2 s)/(rad cycle)] associated to excitation process $z(t)$ is derived as:

$$W_z(\omega) = \frac{W_\zeta(n)}{2\pi v} \tag{6.1}$$

where $\omega = 2\pi v n$ [rad/s] is the angular frequency corresponding to wavelength λ and $W_\zeta(n)$ is the one-sided wave number spectrum of the road.

Another possible normalisation expresses $W_z(\omega)$ as a function of frequency and defines a one-sided spectrum $W_z(f)$ [m^2 s/cycle] as:

$$W_z(f) = \frac{W_z(\omega)}{2\pi} = \frac{W_\zeta(n)}{v} \tag{6.2}$$

Once the road wave number spectrum, $W_\zeta(n)$, and the vehicle speed are known, the spectral density of the excitation is computed by means of Eqs. (6.1) or (6.2).

As far as a certain road section is concerned, its wave number spectrum $W_\zeta(n)$ is fixed, in the sense that it does not vary with the vehicle speed, while the angular frequency spectrum $W_z(\omega)$ (or $W_z(f)$) depends explicitly on the vehicle speed.

In literature, one can find several different explicit expressions for $W_\zeta(n)$ spectrum, see for example [Gobbi and Mastinu 1998, Gobbi and Mastinu 2000, ISO standard, Dodds and Robson 1973]. In our analysis we will adopt the description proposed by Dodds and Robson [Dodds and Robson 1973], which gives the one-sided spectral density (see Figure 6.1):

$$
W_\zeta(n) =
\begin{cases}
W_\zeta(n_0)\left(\dfrac{n}{n_0}\right)^{-b_1}, & n \leq n_0 \\[3mm]
W_\zeta(n_0)\left(\dfrac{n}{n_0}\right)^{-b_2}, & n \geq n_0
\end{cases}
\tag{6.3}
$$

The quantity n, as said before, is the wave number, while $W_\zeta(n_0)$ [m^3/cycle] is a roughness coefficient and represents the value of the spectral density at the discontinuity n_0, which has approximately a constant value of $1/2\pi$ cycle/sec (see Dodds and Robson 1973]. Spectral density $W_\zeta(n)$ needs special attention, since according to its definition, it is not integrable towards high wave numbers; thus, a cut-off wave number n_{max} has to be assumed.

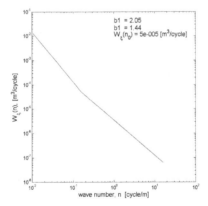

Figure 6.1: Single-sided spectral density $W_\zeta(n)$.

A classification of road profiles, based on coefficients appearing in Eq. (6.3) and proposed in [Dodds and Robson], is reported in Table 6.1.

Table 6.1: Classification of road profiles (see [Dodds and Robson 1973])..

Road class		$W_\varsigma(n_0)*$ range [m³/cycle]	b_1 Mean	b_1 Standard deviation	b_2 Mean	b_2 Standard deviation
Motorways	Very good	2 – 8	1.945	0.464	1.360	0.221
	Good	8 – 32				
Principal roads	Very good	2 – 8	2.05	0.487	1.440	0.266
	Good	8 – 32				
	Average	32 – 128				
	Poor	128 – 512				
Minor roads	Average	32 – 128	2.28	0.534	1.428	0.263
	Poor	128 – 512				
	Very poor	512 – 2048				

6.3. THE CAR-QUARTER MODEL

A simple model, supposed to represent a first approximation of a car-quarter vehicle, is considered (see Figure 6.2). The sprung mass m_2 is assumed to model the quarter vehicle body, with the suspension represented by a spring – damping system (spring stiffness k_2 and damping coefficient c_2). The tyre has a mass m_1, with stiffness represented by k_1. Tyre damping effect is included into the damping coefficient c_1. It is assumed that all springs and dampers are linear.

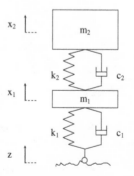

Figure 6.2: The car-quarter model.

When the model moves on the road profile $\zeta(x)$ in x direction at a constant speed v, at time t it is excited by the road vertical displacement $z(t) = \zeta(vt)$, implicitly assuming that the initial position is $x = 0$ at $t = 0$.

The governing equations of the system are ([Sun 2001, Sun and Kennedy 2002]):

$$\begin{cases} m_1 \ddot{x}_1 + c_2(\dot{x}_1 - \dot{x}_2) + c_1(\dot{x}_1 - \dot{z}) + k_2(x_1 - x_2) + k_1(x_1 - z) = 0 \\ \\ m_2 \ddot{x}_2 - c_2(\dot{x}_1 - \dot{x}_2) - k_2(x_1 - x_2) = 0 \end{cases} \tag{6.4}$$

where $x_1(t)$ and $x_2(t)$ are the absolute displacements of the two masses m_1 and m_2, respectively.

Let $y_1(t) = z(t) - x_1(t)$ and $y_2(t) = x_1(t) - x_2(t)$ be the relative displacements between the two masses. Substituting in Eq. (6.4) and taking Fourier transform of both sides, we obtain the harmonic transfer functions, $Y_1(\omega)$ and $Y_2(\omega)$, relating the road vertical profile $z(t)$ to relative displacements $y_1(t)$ and $y_2(t)$:

$$\begin{bmatrix} \omega^2 - i\dfrac{\omega c_1}{m_1} - \dfrac{k_1}{m_1} & i\dfrac{\omega c_2}{m_1} + \dfrac{k_2}{m_1} \\ \\ \omega^2 & \omega^2 - i\dfrac{\omega c_2}{m_2} - \dfrac{k_2}{m_2} \end{bmatrix} \begin{bmatrix} Y_1(\omega) \\ \\ Y_2(\omega) \end{bmatrix} = \begin{bmatrix} \omega^2 \\ \\ \omega^2 \end{bmatrix} \tag{6.5}$$

We are interested in the force $F_1(t)$ acting between the road and the wheel and in the force $F_2(t)$ acting on the suspension system (i.e. the force transferred between the two masses):

$$F_1(t) = k_1 y_1(t) + c_1 \dot{y}_1(t)$$
$$\tag{6.6}$$
$$F_2(t) = k_2 y_2(t) + c_2 \dot{y}_2(t)$$

By taking Fourier transforms of Eq. (6.6), we can define the harmonic transfer functions, $H_1(\omega)$ and $H_2(\omega)$, relating the road vertical profile $z(t)$ to forces $F_1(t)$ and $F_2(t)$:

$$H_1(\omega) = (k_1 + i\omega c_1)Y_1$$
$$\tag{6.7}$$
$$H_2(\omega) = (k_2 + 1\omega c_2)Y_2$$

In a linear analysis, the spectral densities of the two forces, $W_1(\omega)$ and $W_2(\omega)$, are calculated from the input spectral density as [Lutes and Sarkani 2004]:

$$W_i(\omega) = |H_i(\omega)|^2 W_Z(\omega) \qquad i = 1, 2 \tag{6.8}$$

111

where $\left|H_1(\omega)\right|^2$ and $\left|H_2(\omega)\right|^2$, called the load transfer functions, are depicted in Figure 6.3 using parameter values as given in Table 6.3.

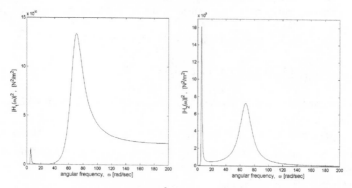

Figure 6.3: Load transfer functions $\left|H_1(\omega)\right|^2$ and $\left|H_2(\omega)\right|^2$ calculated using parameter values given in Table 6.2.

The one-sided spectral densities $W_1(\omega)$ and $W_2(\omega)$ are represented in Figure 6.4, by assuming the parameter values of Table 6.2 and Table 6.3.

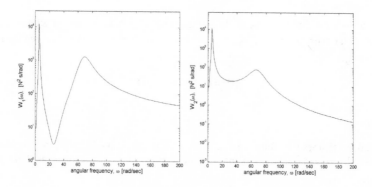

Figure 6.4: One-sided spectral densities $W_1(\omega)$ and $W_2(\omega)$.

6.4. SIMULATIONS AND RESULTS

The numerical simulations define a reference condition for both the values of model parameters and parameters of the spectral density of the road. The parameter values in Table 6.3 define a reference model, which is assumed to move at constant speed

112

$v = 60$ [km/h] on an average principal road. The parameter values of the spectral density $W_\zeta(n)$ of the road are collected in Table 6.2.

Furthermore, the model is assumed to travel along a distance $L = 3000$ [m]. Since distance L represents the maximum wavelength that can be encountered in the road, the corresponding minimum wave number in the wave spectrum is $n_{min} = 1/L$, while a cut-off wave number equal to $n_{max} = 100/2\pi$ [cycles/m] is also assumed.

Table 6.2: Values of the reference road.

Road type	$W_\zeta(n_0)$ [m³/cycle]	b_1	b_2
Principal road (average)	$50 \cdot 10^{-6}$	2.05	1.44

For a given set of model parameters, road spectral density and vehicle speed, the spectral densities $W_1(\omega)$ and $W_2(\omega)$ of the two forces $F_1(t)$ and $F_2(t)$ are firstly computed, as in Eqs. (6.1) and (6.8). Knowledge of these two force spectral densities allows one to simulate a Gaussian time history and to compute the fatigue damage intensity by means of the rainflow count and the Palmgren-Miner rule. At the same time, one can use spectral methods for estimating fatigue damage (under the rainflow count and the linear rule) directly from $W_1(\omega)$ and $W_2(\omega)$.

Therefore, for each set of parameters (i.e. for a given $W_1(\omega)$ and $W_2(\omega)$ pair), we can compare a fatigue damage intensity calculated in time-domain with its estimation from frequency-domain, for both forces $F_1(t)$ and $F_2(t)$.

Table 6.3: Parameter values of reference vehicle.

Model parameter	Value		Description
m_1	30	kg	sprung mass
m_2	500	kg	un-sprung mass
k_1	120000	N/m	tyre stiffness
k_2	20000	N/m	suspension stiffness
c_1	300	N s/m	tyre damping
c_2	3000	N s/m	suspension damping

At this point, an additional aspect to investigate is the effect produced on fatigue damage by changing the value of one model parameter (e.g. suspension stiffness or damping) at a time. This allows us to investigate what influence would be expected on rainflow fatigue damage by changing model parameters.

When a new parameter value is set, response spectral densities $W_1(\omega)$ and $W_2(\omega)$ are computed again; new simulated time histories are rainflow counted and the fatigue damage intensity under the linear rule is compared with its estimation computed in frequency-domain.

113

In our calculations, we assume that the rainflow damage intensity, say $\overline{D}_0$, relative to the reference condition (i.e. parameter values given in Table 6.2 and Table 6.3), and computed by the direct method on simulated time histories, is taken to be equal to unity. Consequently, all results are dimensionless.

Amongst all possible frequency-domain approaches, we have considered the Wirsching-Light method [Wirsching and Light 1980], the Dirlik method [Dirlik 1985] and the TB method illustrated in Chapter 4; in addition, the Sakai-Okamura method (see Chapter 5) is also applied.

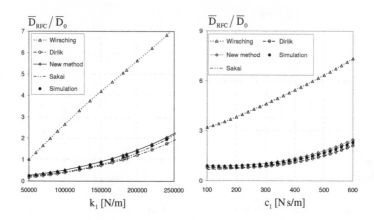

Figure 6.5: Effect of stiffness k_1 and suspension damping c_1 on fatigue damage produced on the vehicle wheel by $F_1(t)$.

114

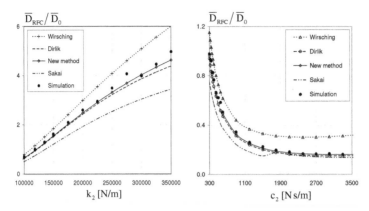

Figure 6.6: Effect of tyre stiffness k_2 and tyre damping c_2 on fatigue damage produced on the suspension by $F_2(t)$.

In Figure 6.5 and Figure 6.6 we illustrate the damage comparison amongst all mentioned methods and the results of damage calculations from simulated time histories. Figure 6.5 refers to the damage produced by the force $F_1(t)$, while Figure 6.6 refers to force $F_2(t)$.

Two main considerations can be drawn from such figures. The first is about the accuracy of each spectral method in predicting the fatigue damage; the second concerns the ability to predict the influence of a change of a chosen parameter model value on the fatigue damage.

As can be seen from the figures, the Wirsching-Light approach is never accurate nor reliable, since it always tends to predict too conservative damage values and also it sometimes overestimates the effect on damage produced by changing the value of a parameter.

On the other hand, the remaining methods seem to be good enough to be used in predicting fatigue damage. The only difference lies in the fact that, while the Sakai-Okamura method strictly applies to bimodal random processes only, the other two approaches (i.e. Dirlik and the new method) are applicable to a wider class of processes.

115

Chapter 7

FATIGUE ANALYSIS OF NON-GAUSSIAN RANDOM LOADINGS

7.1. INTRODUCTION

Chapter 3 showed that that when a mechanical component or structure is subjected to irregular loadings, the fatigue assessment procedure may adopt the concept of stationary random process as a model for the load.

The general assumption is that the random process is Gaussian and, under this hypothesis, we studied some theoretical aspects of the procedure of fatigue damage assessment. In particular, we analysed some methods capable, for a Gaussian process, to explicitly relate the distribution of cycle counted by a given counting procedure (e.g. the rainflow count) to the spectral density of the process, the fatigue damage being subsequently computed by adopting a suitable damage accumulation rule (e.g. Palmgren-Miner linear damage rule).

More specifically, we pointed out how the distribution of cycle counted by the rainflow count is linked to some spectral properties of Gaussian process and how fatigue damage under the Palmgren-Miner rule could be easily computed by integration of the cycle distribution.

We analysed several spectral methods developed for the fatigue analysis of Gaussian random loadings, as the TB method [Tovo 2002, Benasciutti and Tovo 2005a, 2006a]. By means of numerical simulations, we showed how this method is as accurate as the Dirlik method, which is recognised as the most accurate procedure for damage estimation [Bouyssy 1993].

All methods reviewed in Chapter 4 are strictly applicable to stationary random processes, which are also Gaussian. However, a real irregular loading modelled as a stationary process seldom verifies the Gaussian hypothesis. This fact can be due, for example, to a non-linear structural behaviour, to a non-Gaussian excitation (e.g. wind or wave loads), or both. Consequently, in applications including non-Gaussian effects the application to non-Gaussian loadings of spectral methods developed for Gaussian processes can sometimes significantly underestimate the fatigue damage contribution of the highest cycles [Sarkani et al. 1994].

Therefore, in applications concerning non-Gaussian loadings a further theoretical development of the existing Gaussian spectral methods is clearly necessary.

Some works presented in the literature were specifically studied to account for the non-normal behaviour of a stationary process. Nevertheless, they provide analytical estimation formulas which are based on the extension of the narrow-band approximation, i.e. they are restricted to narrow-band spectra only. Moreover, they describe the non-normal character of the load by using only the kurtosis coefficient [Winterstein 1985, Winterstein 1988, Winterstein et al. 1994, Kihl et al 1995]. These cited methods are reviewed at the end of this Chapter.

On the contrary, the demand of methods applicable to broad-band processes that also include a complete description of non-Gaussian effects (e.g. by also including skewness value) is clearly evident.

In a recent work the problem is solved by the definition of a suitable correction factor including both bandwidth and non-normal effects, the latter depending on both skewness and kurtosis coefficients [Yu et al. 2004]. Unfortunately, the method develops only a numerically based approach and does not give any closed-form solution. Furthermore, it is limited to simulated processes and completely disregards possible applications involving real stress time histories.

On the contrary, the method presented in this Chapter tries to include in the damage assessment procedure the non-Gaussian character of a real load in a more complete way. The method is a further extension of the TB method, developed for broad-band Gaussian processes [Tovo 2002, Benasciutti and Tovo 2005a, 2006a] and it is based on a fully-developed theoretical framework that includes the non-normal behaviour of the load (in terms of skewness and kurtosis). In addition, it is specifically developed to apply to broad-band non-Gaussian processes. Since from results on real load histories (see Chapter 8) it seems to be very flexible and sufficiently precise, in our opinion and based of our survey of the existing literature, it appears as a new original contribution in this field.

In this Chapter we will briefly review the main aspects of the cycle distribution and fatigue damage assessment procedure in non-Gaussian loadings, in conjunction with some schemes for modelling non-Gaussian random processes.

7.2. CYCLE DISTRIBUTION IN GAUSSIAN RANDOM LOADINGS

According to the conclusions drawn at the end of Chapter 4, and based on the comparison of results from numerical simulations, only the Dirlik method and the TB method are the most accurate methods, amongst all techniques developed for estimating the rainflow cycle distribution and the fatigue damage,.

However, we further noticed that only the TB method, since capable of estimating explicitly the distribution of rainflow cycles, $h_{RFC}(x_p, x_v)$, as a function of peak x_p and valley x_v levels, is the only one method that can be further extended to include non-Gaussian effects into the fatigue damage assessment procedure. In this Chapter we will see the main theoretical aspects of the framework that support this assertion.

It is our intention to recall here some parts of the theoretical development of the TB method, since a slightly different notation will be used in the remaining part of this Chapter, where the method will be further developed to the non-Gaussian case. The complete treatment for the case of Gaussian random processes can be found in Chapter 4.

Let $X(t)$ be a Gaussian random process and let $h_X^G(x_p, x_v)$ be the joint probability density function, representing the distribution of cycles counted by a given counting procedure, as function of peak x_p and valley x_v levels. The cumulative distribution function $H_X^G(x_p, x_v)$ is defined as:

$$H_X^G(x_p, x_v) = \int_{-\infty}^{x_p} \int_{-\infty}^{x_v} h_X^G(x, y) \, dx \, dy \tag{7.1}$$

and it gives the probability to count a cycle with peak lower or equal to level x_p and valley lower or equal to level x_v. The subscript and superscript remind us that we are dealing with the distribution of cycles counted in process $X(t)$ and that process $X(t)$ is Gaussian.

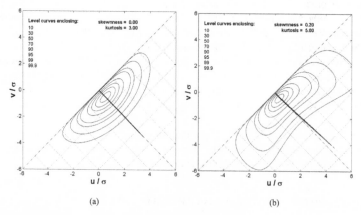

(a)   (b)

Figure 7.1: Comparison of rainflow cycle distribution $h_{RFC}(x_p, x_v)$ as a function of peak x_p and valley x_v for two processes with same variance $\sigma^2 = 1$: (a) joint density $h_{RFC}^G(x_p, x_v)$ for a Gaussian process; (b) joint density $h_{RFC}^{nG}(x_p, x_v)$ for a non-Gaussian process with $\gamma_3 = 0.2$ and $\gamma_4 = 5$. Representation of Dirac delta function is qualitative.

Referring in particular to cycles counted by the rainflow method, we will write $h_{X,RFC}^G(x_p, x_v)$ for the joint density and $H_{X,RFC}^G(x_p, x_v)$ for its distribution. The rainflow distribution is estimated by the linear combination (see Chapter 4):

$$h_{X,RFC}^{G}(x_p, x_v) = b\, h_{X,LCC}^{G}(x_p, x_v) \; + \; (1-b)\, h_{X,RC}^{G} \qquad (7.2)$$

where b is a weighting coefficient depending on the spectral density of the process, $h_{X,LCC}^{G}(x_p, x_v)$ and $h_{X,RC}^{G}(x_p, x_v)$ are the cycle distributions for the level-crossing count and the range count, respectively. Their explicit expressions are given in Chapter 4.

As already noticed, the true correlation between b and the process spectral density is known only in terms of approximate relations. Since the distribution for the range count is given by an approximate formula too, Eq. (7.2) must be considered as an approximation. In Figure 7.1(a) we show the rainflow joint density function $h_{X,RFC}^{G}(x_p, x_v)$, calculated as in Eq. (7.2), for a Gaussian process having variance $\sigma_X^2 = 1$.

Fatigue damage under the Palmgren-Miner damage rule is computed from the marginal amplitude distribution $p_{X,RFC}^{G}(s)$ derived from $h_{X,RFC}^{G}(x_p, x_v)$ joint density, see Chapter 3.

Since Eq. (7.2) is a linear relation, a similar expression holds also for other distributions, as for example the cumulative distribution:

$$H_{X,RFC}^{G}(x_p, x_v) = b\, H_{LCC}^{G}(x_p, x_v) \; + \; (1-b)\, H_{RC}^{G}(x_p, x_v) \qquad (7.3)$$

where $H_{LCC}^{G}(x_p, x_v)$ and $H_{RC}^{G}(x_p, x_v)$ are the cumulative distribution functions associated to the joint densities $h_{X,LCC}^{G}(x_p, x_v)$ and $h_{X,RC}^{G}(x_p, x_v)$ for the level-crossing count and range count, as in Eq. (7.1).

More precisely, integration of the analytical expression for $h_{X,LCC}^{G}(x_p, x_v)$ given in Chapter 4 gives the cumulative distribution for the level-crossing count:

$$H_{X,LCC}^{G}(x_p, x_v) =$$

$$\left[P_p(x_v) \; + \; \alpha_2 \left(e^{-\frac{x_v^2}{2\lambda_0}} - e^{-\frac{x_p^2}{2\lambda_0}} \right) \mathbf{I}(x_p + x_v) \right] \mathbf{I}(-x_v) \; + \qquad (7.4)$$

$$+ \left[P_v(x_v) \; - \; \alpha_2\, e^{-\frac{x_p^2}{2\lambda_0}} \right] \mathbf{I}(x_v)$$

$\mathbf{I}(x) = 0$ being an indicator function ($\mathbf{I}(x) = 1$ if $x \geq 0$, elsewhere $\mathbf{I}(x) = 0$) and $P_p(x)$ and $P_v(x)$ the peak and valley cumulative distribution functions, see Chapter 3. Similarly, integration of $h_{X,RC}^{G}(x_p, x_v)$ gives the distribution for the range counting:

$$H_{RC}^{G}(x_p, x_v) =$$

$$\Phi\left(\frac{x_v}{\sqrt{\lambda_0(1-\alpha_2^2)}}\right) - \alpha_2\, e^{-\frac{x_p^2}{2\lambda_0}}\, \Phi\left(\frac{x_v - x_p\,(1-2\alpha_2^2)}{2\,\alpha_2\,\sqrt{\lambda_0(1-\alpha_2^2)}}\right) + \tag{7.5}$$

$$+ \alpha_2\, e^{-\frac{x_v^2}{2\lambda_0}}\left[\Phi\left(\frac{x_p - x_v\,(1-2\alpha_2^2)}{2\,\alpha_2\,\sqrt{\lambda_0(1-\alpha_2^2)}}\right) - \Phi\left(\frac{\alpha_2\, x_v}{\sqrt{\lambda_0(1-\alpha_2^2)}}\right)\right]$$

where $\Phi(\cdot)$ is the standard normal distribution function. All densities and cumulative distributions are defined only for $x_p - x_v \geq 0$.

7.3. Cycle distribution in non-Gaussian random loadings

The previous method (summarised in Eqs. (7.2) and (7.3)) only applies to stationary random processes, which are also Gaussian. However, in modelling a measured stress response as a random process, we observe that the Gaussian hypothesis is seldom verified, due for example to a non-linear behaviour of the system or to a non-Gaussian excitation (e.g. wind and wave loads), or both. As an example, Figure 7.2 shows a normal probability paper for load values measured experimentally: the marked deviation from the straight line (i.e. the Gaussian assumption) points out that data are not Gaussian.

In this case, the inappropriate use of the methods valid under the Gaussian hypothesis can lead to incorrect estimations, which underestimate the frequency of high level cycles and their related fatigue damage contribution [Winterstein 1988, Kihl et al. 1995].

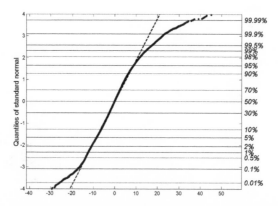

Figure 7.2: Normal probability plot for a time history $z(t)$ taken from a non-Gaussian random process $Z(t)$. The straight line indicates the Gaussian behaviour; any deviation from that line evidences a non-Gaussian character of the measured load.

Consequently, it would be of great importance if the TB method summarised above was able to also account for non-Gaussian effects of real loads, when present. In this section, we shall see how, in order to do this, a given non-Gaussian process can be related to an underlying Gaussian one through an appropriate transformation, which establishes a link between their corresponding rainflow cycle distributions.

We know that a random process $X(t)$ is Gaussian if its values follow a normal distribution around its mean value μ_X:

$$p_X(x) = \frac{1}{\sqrt{2\pi}\,\sigma_X} e^{-\frac{1}{2}\left(\frac{x-\mu_X}{\sigma_X}\right)^2}$$

(7.6)

where the distribution width around the mean is measured by the variance σ_X^2.

On the contrary, a non-Gaussian process $Z(t)$ deviates from the normal distribution; this deviation is usually characterised by the skewness and the kurtosis, defined as:

$$\gamma_3 = \frac{E\left[(Z-\mu_Z)^3\right]}{\sigma_Z^3} \qquad \text{skewness}$$

$$\gamma_4 = \frac{E\left[(Z-\mu_Z)^4\right]}{\sigma_Z^4} \qquad \text{kurtosis}$$

(7.7)

The skewness identifies the degree of asymmetry of a non-Gaussian distribution, i.e. for a distribution symmetric around its mean value $\gamma_3 = 0$. The kurtosis quantifies the contribution of the tails of a non-normal density: for a wider than Gaussian distribution tails (leptokurtic) kurtosis is greater then three, whereas the converse (i.e. kurtosis less then three) holds for a density having less probability mass in the tails (platykurtic). Obviously, a Gaussian process has $\gamma_3 = \gamma_4 - 3 = 0$.

Given a Gaussian process $X(t)$, we introduce a non-linear transformation $G(\cdot)$, by which a non-normal process $Z(t)$ can be generated according to:

$$Z(t) = G(X(t))$$

(7.8)

As a consequence, the inverse transformation $g(\cdot) = G^{-1}(\cdot)$ can be used to take back a non-Gaussian process $Z(t)$ to a corresponding underlying Gaussian one, $X(t) = g(Z(t))$. Examples of such a transformation are depicted in Figure 7.5 and Figure 7.6. Figure below shows an example of a non-Gaussian process with its underlying Gaussian process.

Note that choosing the transformation $G(\cdot)$ as a monotonically increasing function allow the process $X(t)$ and its corresponding transformed $Z(t)$ to have mean upcrossings, peaks or valleys at the same instants of time, respectively; then, both processes will possess the same irregularity factor IF.

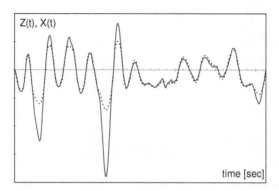

Figure 7.3: Non-Gaussian process $Z(t)$ (continuous line) and its underlying Gaussian process $X(t)$ (dotted line).

Referring in particular to the correspondence between peaks and valleys, if $X(t)$ has a peak at level x_p at time t_i, $Z(t)$ will have a corresponding peak at level $z_p = G(x_p(t_i))$ at the same instant of time. Similarly, a valley x_v is transformed in $z_v = G(x_v(t_i))$.

Furthermore, by imposing the transformation as being monotonically non-decreasing allow us to also preserve the relative position amongst peaks and valleys. For example, if $x_p(t_1) > x_p(t_2)$, then it will be $z_p(t_1) > z_p(t_2)$, and the same for valleys. Consequently, since in a random process the rainflow counting extracts cycles by pairing peaks and valleys based on their relative positions and time sequence (so forming complete cycles or half-cycles), rainflow counted cycles in the $X(t)$ and transformed $Z(t)$ process will be counted at the same instants of time, and they will be formed by coupling pairs of corresponding peaks and valleys. In other words, the transformation, without changing the number and the sequence of rainflow counted cycles, modifies instead their amplitudes and mean values by increasing or decreasing their maximum and minimum values.

This fact then enables us to establish a fundamental relation between a rainflow cycle counted in the $X(t)$ process and its correspondent cycle in the transformed $Z(t)$ process: given a cycle in process $X(t)$ having peak $x_p(t_i)$ and valley $x_v(t_j)$, we can associate to it a corresponding cycle in process $Z(t)$, having peak $z_p(t_i)$ and valley $z_v(t_j)$. Denoting (x_p, x_v) as a cycle, the corresponding transformed cycle will be (z_p, z_v), with:

$$\text{if } (x_p, x_v) \quad \Rightarrow \quad (z_p, z_v) = \vec{G}((x_p, x_v)) = (G(x_p), G(x_v)) \qquad (7.9)$$

The direct transformation has been applied separately to peak and valley. Since rainflow counted cycles in a random process represent a set of random variables, the previous relation also establishes a link between two pairs of random variables. Consequently, the following important relation that links the rainflow cumulative cycle distributions for two processes $X(t)$ and $Z(t)$ can be stated:

$$H_{Z,\text{RFC}}(z_p, z_v) = H_{X,\text{RFC}}\big(g(z_p), g(z_v)\big) = H_{X,\text{RFC}}(x_p, x_v) \tag{7.10}$$

In this case the inverse transformation $g(\cdot)$ has been used. This relation summarises the idea that the probability of counting a cycle in the $X(t)$ process with peak lower than or equal to x_p and valley lower than or equal to x_v is the same of counting a cycle in the $Z(t)$ process with peak lower than or equal to z_p and valley lower than or equal to z_v, given by Eq. (7.9). This is true for a non-decreasing $G(\cdot)$ function, and not necessarily implying that $X(t)$ and $Z(t)$ must be Gaussian and non-Gaussian, respectively.

Obviously, for the case in which the applied transformation $G(\cdot)$ is used to link a given non-normal process $Z(t)$ to its underlying Gaussian one, $X(t)$, the previous relation further specializes: the non-Gaussian rainflow cumulative cycle distribution, say $H_{Z,\text{RFC}}^{\text{nG}}(z_p, z_v)$, is linked to the corresponding rainflow cumulative cycle distribution in the Gaussian domain, say $H_{X,\text{RFC}}^{\text{G}}(x_p, x_v)$:

$$H_{Z,\text{RFC}}^{\text{nG}}(z_p, z_v) = H_{X,\text{RFC}}^{\text{G}}\big(g(z_p), g(z_v)\big) = H_{X,\text{RFC}}^{\text{G}}(x_p, x_v) \tag{7.11}$$

At this point, we know how to estimate the rainflow cycle distribution for the Gaussian process, namely by using a linear combination as in Eq. (7.3). This allows us to estimate the rainflow cycle distribution for the non-Gaussian process, $H_{Z,\text{RFC}}^{\text{nG}}(z_p, z_v)$.

The use of Eq. (7.3) allows one to estimate in a non-Gaussian process $Z(t)$ the rainflow cycle cumulative distribution, $H_{Z,\text{RFC}}^{\text{nG}}(z_p, z_v)$, from which the corresponding joint density $h_{Z,\text{RFC}}^{\text{nG}}(z_p, z_v)$ can be easily obtained by differentiation. In Figure 7.1(b) the rainflow cycle distribution for a non-Gaussian process with $\sigma_Z^2 = 1$, having $\gamma_3 = 0.2$ and $\gamma_4 = 5$, is shown .Comparison with the cycle distribution for a Gaussian process with same variance, depicted in Figure 7.1(a), clarifies that in the non-Gaussian case higher probability is associated to larger cycles, with respect to Gaussian distribution.

Finally, the previous theoretical development has underlined an important fact: a necessary condition for estimating the rainflow non-Gaussian distribution from the corresponding Gaussian one is the knowledge of the whole distribution $h_{\text{RFC}}(x_p, x_v)$ (or equivalently $H_{\text{RFC}}(x_p, x_v)$) as a function of peak and valley levels, and not simply the knowledge of marginal distribution of amplitudes alone. In this sense, other methods (as that developed by Dirlik), able to estimate only the marginal amplitude distribution of rainflow cycles in a Gaussian process, even if showing satisfactory results, cannot be further extended to the non-Gaussian case.

7.4. DEFINITION OF THE TRANSFORMATION

In literature, there are several parametric formulas that can be used to define the transformation $G(\cdot)$. That proposed by Ochi is a monotonic exponential function [Ochi and Ahn 1994], Winterstein's model is a monotonic cubic Hermite polynomial [Winterstein 1985 and 1988, Winterstein et al. 1994], while the transformation proposed by Sarkani et al. uses a power-law model [Sarkani et al. 1994, Kihl et al. 1995]. Coefficients of the transformation are functions of the skewness and kurtosis of the non-Gaussian process and must be estimated from real data. In alternative, we can use a non-parametric definition of the transformation, as that recently proposed by Rychlik et al. [Rychlik et al. 1997] to model ocean-wave data.

In the next paragraphs we will review in some detail all cited models. In particular, we note here that the model proposed by Winterstein is adopted in our analysis, since it is shown to provide good accuracy in representing a wide range of non-linear behaviours.

All models establish a functional relationship between $X(t)$ and $Z(t)$ processes in terms of either skewness, γ_3 or kurtosis, γ_4, (or both), of the non-Gaussian process $Z(t)$. From now on, the time argument t is omitted for sake of clarity, even if all relations should be understood as applied at each instant in time.

7.4.1. Power-law model [Sarkani et al. 1994]

This simple model defines the transformation as a power-law function, which quantifies the deviation from the Gaussian hypothesis only in terms of the kurtosis (i.e. it implicitly applies to symmetric non-Gaussian processes).

Let us assume that $X(t)$ and $Z(t)$ are the normal and the non-normal random processes, both having a zero mean value; the $G(\cdot)$ function relating them as in Eq. (7.8) is supposed to have the following form:

$$Z = G(X) = \frac{X + \beta\left(\mathrm{sgn}(X)\right)\left(\mid X \mid^n\right)}{D} \tag{7.12}$$

in which $\mathrm{sgn}(\cdot)$ is the signum function ($\mathrm{sgn}(x) = 1$ for $x > 0$, 0 for $x = 0$ and -1 for $x < 0$) and β and n are parameters controlling the degree of non-normality of the non-Gaussian process. When the two parameters, β and n, are confined to positive values, they result in a monotonically increasing function. The function is normalised by a constant D such that the non-normal process $Z(t)$ has the same variance as the normal process $X(t)$ (i.e. $\sigma_X^2 = \sigma_Z^2$):

$$D = \sqrt{1 + \frac{2^{(n+1)/2}\, n\, \Gamma\!\left(\dfrac{n}{2}\right)\sigma_X^{n-1}}{\sqrt{\pi}}\,\beta + \frac{2^n\, \Gamma\!\left(n+\dfrac{1}{2}\right)\sigma_X^{2(n-1)}}{\sqrt{\pi}}\,\beta^2} \tag{7.13}$$

The kurtosis of the non-normal process $Z(t)$ is given by:

124

$$\gamma_4 = \frac{E[Z^4]}{\sigma_Z^4} = \frac{E\left[\left(X + \beta\left(\mathrm{sgn}(X)\right)|X^n|\right)^4\right]}{D^4\,\sigma_Z^4}$$

$$\text{(7.14)}$$

$$= \frac{1}{D^4\,\sigma_X^4}\left(E[X^4] + 4\beta\,E[|X|^{n+3}] + 6\beta^2\,E[X^{2n+2}] +\right.$$
$$\left. + 4\beta^3\,E[|X|^{3n+1}] + \beta^4\,E[X^{4n}]\right)$$

where the moments appearing in equation above can be computed as:

$$E[|X|^m] = \frac{(2\sigma_X^2)^{m/2}}{\sqrt{\pi}}\,\Gamma\!\left(\frac{m+1}{2}\right) \qquad (m \text{ integer}) \qquad \text{(7.15)}$$

Using the kurtosis as a measure of non-normality, a unique transformation can be obtained by specifying the kurtosis, the variance σ_X^2 or σ_Z^2, and either the intensity of non-normality, n, or the coefficient of the non-linear portion, β. Note however, that many different transformations, defined by β and n, can produce the same kurtosis value, as clearly evidenced in Figure 7.4, which refers to a process with $\sigma_X = \sigma_Z = 1$.

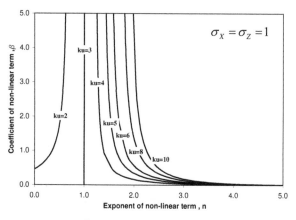

Figure 7.4: Relationship between β and n parameters for the transformation $G(\cdot)$ in the power law model.

Note that if process $Z(t)$ is a platykurtic non-normal process ($\gamma_4 < 3$), it follows that n is less than one, while if it is a leptokurtic process ($\gamma_4 > 3$), n must be greater than one; finally, if $Z(t)$ is Gaussian ($\gamma_4 = 3$), the transformation requires that $\beta = 0$ and $D = 1$.

125

Table 7.1: Values of the transformation parameters of the power-law model for two values of the kurtosis γ_4 (processes have $\sigma_X = \sigma_Z = 1$)

γ_4	n	β	D
2	0.5	1.735	2.527
5	2	0.342	1.563

As an example, in Table 7.1 we report the coefficients for the case $\sigma_X = \sigma_Z = 1$. In the figure below, we sketch the transformation given in Eq. (7.13) for a non-Gaussian process, having kurtosis $\gamma_4 = 5$. Obviously, all previous results concerning the definition of the transformation $G(\cdot)$ can be easily updated for the case of non-zero mean value.

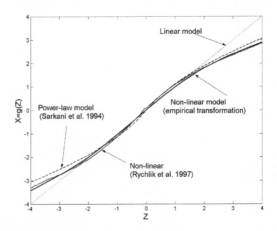

Figure 7.5: Inverse transformations $X(t) = g\big(Z(t) \big)$ for process $Z(t)$ with $\mu_Z = 0$, $\sigma_Z^2 = 1$, $\gamma_3 = 0.5$, $\gamma_4 = 5$. Linear model (dotted line), power-law model (dashed line), non-linear model (solid line), with the empirical transformation.

7.4.2. Ochi and Ahn model (1994)

This is a model where the transformation is chosen to be a monotonically increasing exponential function, calibrated on the basis of the first three moments (i.e. mean, variance and skewness) of the non-Gaussian process $Z(t)$ [Ochi and Ahn 1994, Ochi 1998].

The inverse function $g(\cdot)$ relating the non-Gaussian world to the Gaussian one is defined in the following way [Ochi and Ahn 1994]:

$$X = \frac{1}{\gamma a} \left(1 - e^{-\gamma a Y} \right) \tag{7.16}$$

where X is a Gaussian variable with mean μ_* and variance σ_* (the notation is conform to the original paper) and Y is a non-Gaussian variable with mean $\mu_Y = 0$, variance σ_Y^2 and skewness sk_Y. The magnitude of the coefficient a in previous expression represents the intensity of the non-linear characteristic (i.e. the larger the a value, the stronger the non-linearity).

Parameter γ appearing into the definition is a constant defined on the basis of the mean value of Y:

$$\gamma = \begin{cases} 1.28 & \text{for} \quad Y \geq 0 \\ 3 & \text{for} \quad Y < 0 \end{cases} \tag{7.17}$$

whereas a, μ_* and σ_* are determined as solutions of the non-linear system:

$$\begin{cases} a(\sigma_*^2 + \mu_*^2) + \mu_* = 0 \\ \sigma_*^2 - 2a^2 \sigma_*^4 = \kappa_2 \\ 2a\sigma_*^4(3 - 8a^2\sigma_*^2) = \kappa_3 \end{cases} \tag{7.18}$$

being κ_i the cumulants of the Y variable:

$$\begin{aligned} \kappa_1 &= E[Y] = \mu_Y = 0 \\ \kappa_2 &= E[(Y - \mu_Y)^2] = \sigma_Y^2 \\ \kappa_3 &= E[(Y - \mu_Y)^3] = \sigma_Y^3 \, sk_Y \end{aligned} \tag{7.19}$$

In a more general case, we refer to a non-Gaussian process $Z(t)$ having mean μ_Z, variance σ_Z^2 and skewness sk_Z. Therefore, Eq. (7.16) will be slightly modified by referring to a standard normal variable X_0:

$$X_0 = \frac{X - \mu_*}{\sigma_*} = \left[\frac{1}{\gamma a} \left(1 - e^{-\gamma a \left(\frac{Z - \mu_Z}{\sigma_Z} \right)} \right) - \mu_* \right] \frac{1}{\sigma_*} \tag{7.20}$$

and by introducing the non-Gaussian random variable $Y = (Z - \mu_Z)/\sigma_Z$, which, as can be easily proved, has mean $\mu_Y = 0$, variance $\sigma_Y^2 = 1$ and skewness $sk_Y = sk_Z$ (i.e. the same skewness as the original $Z(t)$ process). Parameters a, μ_* and σ_* in previous equation are still determined as solutions of the non-linear system Eq. (7.18), updated this time with $\kappa_2 = 1$ and $\kappa_3 = sk_Z$ (see routine ochitr in WAFO toolbox); also Eq. (7.17) updates in terms of the mean value μ_Z of the new process $Z(t)$.

An example of the transformation proposed by Ochi and Ahn is depicted in Figure 7.6 for the case of a non-Gaussian process $Z(t)$ with $\mu_Z = 0$, $\sigma_Z^2 = 1$ and $sk_Z = \gamma_3 = 0.5$.

Some comments are of interest. According to the authors, the proposed model is appropriate also for processes with very strong non-linear characteristics (i.e. highly non-Gaussian). The analysis of the transformation reveals that the model depends only on the skewness of the non-normal process (i.e. it is an asymmetric function, see Figure 7.6). This indicates that the model is not good for symmetric non-Gaussian processes. In fact, it was proposed for modelling the non-linear behaviour resulting from sea waves in finite water depth [Ochi and Ahn 1994].

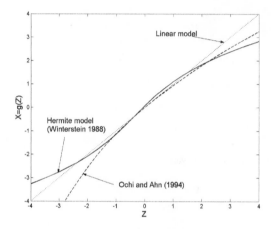

Figure 7.6: Inverse transformations $X(t) = g\left(Z(t) \right)$ for process $Z(t)$ with $\mu_Z = 0$, $\sigma_Z^2 = 1$, $\gamma_3 = 0.5$, $\gamma_4 = 5$. Linear model (dotted line), Hermite model (solid line), Ochi and Ahn model (dashed line).

7.4.3. Hermite model [Winterstein 1985 and 1988, Winterstein et al. 1994]

This model seeks Hermite polynomial approximations for either the transformation $G(\cdot)$ or its inverse, based on moments of the non-Gaussian process. It describes the non-normal behaviour in terms of both skewness and kurtosis.

Models are distinguished for the case of a softening and hardening response process $Z(t)$, which correspond to the case of a leptokurtic and platykurtic.

For a softening response process (i.e. leptokurtic non-Gaussian process $Z(t)$, having $\gamma_4 > 3$), the Winterstein's model defines the direct transformation $G(\cdot)$:

$$Z = \mu_Z + \sigma_Z \, K \left[X_0 + \tilde{h}_3 \left(X_0^2 - 1 \right) + \tilde{h}_4 \left(X_0^3 - 3X_0 \right) \right] \tag{7.21}$$

128

in terms of the standardised normal process $X_0 = (X - \mu_X)/\sigma_X$; the terms $\tilde{h}_3$ and $\tilde{h}_4$ are approximate coefficients related to skewness and kurtosis of the non-Gaussian process, while the K coefficient is a scale factor ensuring that both Gaussian and non-Gaussian processes have the same variance.

The earliest version of this model (first-order model) [Winterstein 1985] assumed a small non-linear (and thus non-normal) behaviour and gave the following coefficients:

$$\tilde{h}_n = h_n \qquad\qquad K = 1 \qquad\qquad (7.22)$$

where h_n are the Hermite moments, with $h_1 = h_2 = 0$ and:

$$h_3 = \frac{\gamma_3}{6} \qquad , \qquad h_4 = \frac{(\gamma_4 - 3)}{24} \qquad\qquad (7.23)$$

Subsequent studies gave results based on a more accurate approximation, by introducing second order terms into the Hermite model (second-order model) [Winterstein 1988]. Coefficients in Eq. (7.21) are precisely:

$$\tilde{h}_4 = \frac{\sqrt{1 + 1.5(\gamma_4 - 3)} - 1}{18} \qquad\qquad \tilde{h}_3 = \frac{\gamma_3}{6(1 + 6\tilde{h}_4)}$$

$$K = \frac{1}{\sqrt{1 + 2\tilde{h}_3^2 + 6\tilde{h}_4^2}} \qquad\qquad (7.24)$$

Furthermore, an alternative version gave the following expressions for coefficients in Eq. (7.21) [Winterstein et al. 1994]:

$$\tilde{h}_3 = \frac{\gamma_3}{6} \left[\frac{1 - 0.15 |\gamma_3| + 0.3\gamma_3^2}{1 + 0.2(\gamma_4 - 3)} \right]$$

$$\qquad\qquad (7.25)$$

$$\tilde{h}_4 = \tilde{h}_{40} \left[1 - \frac{1.43\gamma_3^2}{\gamma_4 - 3} \right]^{1 - 0.1\gamma_4^{0.8}} \qquad\qquad \tilde{h}_{40} = \frac{[1 + 1.25(\gamma_4 - 3)]^{1/3} - 1}{10}$$

Results given in Eq. (7.25) are intended to apply for $3 < \gamma_4 < 15$ and $0 \le \gamma_3^2 < 2(\gamma_4 - 3)/3$, which should include most cases of practical interest. Note that for small non-linearity, Eq. (7.24) is consistent with Eq. (7.22).

An example of $G(\cdot)$ transformation for a non-Gaussian process is depicted in Figure 7.6.

In order to get the inverse function $g(\cdot)$, we have to find solution of Eq. (7.21) for X_0 in terms of Z; by inverting that expression, we obtain the transformation defining the standard normal process X_0 as:

$$X_0 = \left[\sqrt{\xi^2(Z)+c} + \xi(Z)\right]^{1/3} - \left[\sqrt{\xi^2(Z)+c} - \xi(Z)\right]^{1/3} - a \qquad (7.26)$$

in which:

$$\xi(Z) = 1.5b\left(a + \frac{Z-\mu_Z}{K\sigma_Z}\right) - a^3$$

$$a = \frac{\tilde{h}_3}{3\tilde{h}_4} \;\; ; \quad b = \frac{1}{3\tilde{h}_4} \;\; ; \quad c = \left(b - 1 - a^2\right)^3 \qquad (7.27)$$

For an hardening response process (i.e. a platykurtic non-Gaussian process $Z(t)$, having $\gamma_4 < 3$), the inverse transformation $g(\cdot) = G^{-1}(\cdot)$ is given in terms of the standardized process $Z_0 = (Z - \mu_Z)/\sigma_Z$:

$$X_0 = Z_0 - h_3\left(Z_0^2 - 1\right) - h_4\left(Z_0^3 - 3Z_0\right) \qquad (7.28)$$

being $h_3 = \gamma_3/6$ and $h_4 = (\gamma_4 - 3)/24$ the Hermite moments as in Eq. (7.23).

The fundamental requirement for transformation $G(\cdot)$ of being monotonic introduces a limitation on the values of skewness and kurtosis that can be treated by the Hermite model (i.e. not all non-Gaussian processes are allowed).

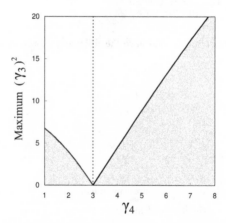

Figure 7.7: A non-Gaussian process having γ_4 and γ_3^2 values falling outside the shadowed area is not treatable by the Hermite model (i.e. it gives a transformation $G(\cdot)$ which is not monotonic).

By imposing that the first derivative of Eq. (7.21) and (7.28) has to be always positive, we may derive a closed-form functional relationship between the skewness and the kurtosis values that can be treated by the model, summarised in Figure 7.7 in terms of the maximum γ_3^2 for each γ_4 value [Winterstein 1988]. This means that a non-

Gaussian process having skewness and kurtosis falling outside the shadowed area (e.g. a kurtosis close to 3, with a high skewness) is not manageable by the Hermite model.

7.4.4. Non-linear model [Rychlik et al. 1997]

The definition of the transformation assumes that the Gaussian process $X(t)$ is normalised, so to have a zero mean value and variances $\sigma_{\dot{X}}^2 = \sigma_{X}^2 = 1$. The level-crossing spectrum, say $v_X(x)$, is given by the well-known Rice's formula (see Chapter 3). Under these circumstances, the level-crossing spectrum for the non-Gaussian process $Z(t)$, say $v_Z(z)$, is obtained as:

$$v_Z(z) = v_X\big(g(z) \big) = \frac{1}{2\pi} \exp\left(-\frac{g(z)^2}{2} \right) \tag{7.29}$$

where $g(\cdot) = G^{-1}(\cdot)$ is the inverse function; note that $v_Z(z)$ has only one local maximum at $z_0 = G(0)$ (i.e. corresponding to the maximum of $v_X(x)$ occurring at $x = 0$).

Then, if the non-Gaussian level-crossing crossing intensity $\mu_Z(z)$ is unimodal (i.e. it has only one maximum), it can be inverted, so giving the following definition of the inverse transformation:

$$g(z) = \begin{cases} \sqrt{-2\ln\big(2\pi v_Z(z)\big)} & \text{if} \quad z \geq z_0 \\[2ex] -\sqrt{-2\ln\big(2\pi v_Z(z)\big)} & \text{if} \quad z < z_0 \end{cases} \tag{7.30}$$

In other words, the knowledge of the non-Gaussian level-crossing spectrum $v_Z(z)$ would give the inverse transformation $g(\cdot)$ through Eq. (7.30). However, the crossing intensity $v_Z(z)$ is usually unknown analytically and has to be estimated from the empirical level-crossing spectrum, say $\hat{v}_Z(z)$.

A procedure is illustrated in [Rychlik et al. 1997] by which the transformation $g(\cdot)$ is estimated from the empirical crossing intensity $\hat{v}_Z(z)$. An example of $g(\cdot)$ transformation is shown in Figure 7.5.

7.5. NARROW-BAND APPROXIMATION IN THE NON-GAUSSIAN CASE

One of the simplest measures of the effect of non-normality on fatigue damage accumulation is the ratio, say η, of the fatigue damage under the non-normal loading $Z(t)$ to that of the normal loading $X(t)$, i.e.

$$\eta = \frac{\overline{D}^{\,nG}}{\overline{D}^{\,G}} \tag{7.31}$$

computed here in terms of expected damage intensities.

Very simple closed-form expressions can be obtained for η index in narrow-band processes. In this case, the damage intensity for the Gaussian case, $\overline{D}^G$, is computed according to the narrow-band approximation:

$$\overline{D}_{NB}^G = v_0 \frac{E\left[s^k\right]}{C} \tag{7.32}$$

which assumes $E[s^k]/C$ as the mean damage per cycle and v_0 (i.e. the mean upcrossing intensity) as the number of cycles counted in the unit time interval. In a narrowband process, each cycle is approximately symmetric about the mean, with amplitude $s = (x_p - x_v)/2$ which is Rayleigh distributed, so that formula writes explicitly as (see Chapter 4):

$$\overline{D}_{NB}^G = \frac{v_0}{C} \left(\sqrt{2\,\sigma_x^2}\right)^k \Gamma\left(1 + \frac{k}{2}\right) \tag{7.33}$$

On the other hand, choosing the $G(\cdot)$ function to be monotonic preserves the mean upcrossing intensity v_0 between the Gaussian loading $X(t)$ and its corresponding non-Gaussian $Z(t)$, and also assures a one-to-one correspondence between peaks and valleys, so that amplitude $s = (x_p - x_v)/2$ transforms into $(z_p - z_v)/2 = (G(x_p) - G(x_v))/2$.

Consequently, the narrow-band approximation formula for the non-normal process can be written as:

$$\overline{D}_{NB}^{nG} = v_0 \frac{E\left[\left(\dfrac{G(x_p) - G(x_v)}{2}\right)^k\right]}{C} \tag{7.34}$$

Obviously, the possibility to have a closed-form expression for the non-normal damage intensity $\overline{D}_{NB}^{nG}$ strictly depends on the form assumed for the transformation $G(\cdot)$. In the following, we will present results by Winterstein [Winterstein 1985 and 1988] and by Sarkani et al. [Sarkani et al. 1994].

7.5.1. Results based on the power-law model [Sarkani et al. 1994]

Referring to the power-law model given in Eq. (7.12), the mean damage intensity given in Eq. (7.34) writes explicitly as:

$$\overline{D}_{NB}^{nG} = v_0 \frac{E\left\{\left[\dfrac{s + \beta\, s^n}{D}\right]^k\right\}}{C} \tag{7.35}$$

where s (i.e. the Gaussian amplitude) is a Rayleigh distributed variable with variance σ_X^2, so that equation above becomes:

$$\overline{D}_{NB}^{nG} = \frac{v_0}{C} \int_0^\infty \frac{s(s + \beta s^n)}{D^k \sigma_X^2} e^{-\frac{s^2}{2\sigma_X^2}} \, ds \qquad (7.36)$$

The non-normal damage intensity $\overline{D}_{NB}^{nG}$ is obtained by numerically integrating Eq. (7.36). In order to simplify this calculation, we propose here an alternative original version of the above result.

By using the definition of binomial coefficients, the exponential terms appearing in the numerator of Eq. (7.35) can be alternatively written as:

$$\left(s + \beta s^n\right)^k = \sum_{p=0}^k \binom{k}{p} s^{k-p} \left(\beta s^n\right)^p \qquad (7.37)$$

Therefore, the non-normal damage intensity, Eq. (7.35), can be expressed as:

$$\overline{D}_{NB}^{nG} = \frac{v_0}{C D^k} E\left[\left(s + \beta s^n\right)^k\right] = \frac{v_0}{C D^k} \sum_{p=0}^k \binom{k}{p} \beta^p E\left[s^{k+n\,p-p}\right] \qquad (7.38)$$

being $E[s^{k+n\,p-p}]$ the moment of the Rayleigh distributed random variable s. From Eqs. (7.33) and (7.38), the η index can be computed explicitly as:

$$\eta = \frac{1}{\Gamma\left(1 + k/2\right) D^k} \sum_{p=0}^k \binom{k}{p} \beta^p \, 2^{p(n-1)/2} \, \sigma_X^{p(n-1)} \, \Gamma\left(1 + \frac{k + n(p-1)}{2}\right) \qquad (7.39)$$

once we know the parameters n, β and D defining the transformation $G(\cdot)$ in Eq. (7.12), and the standard deviation σ_X of both the Gaussian and non-Gaussian process (since $\sigma_X = \sigma_Z$). Respect to Eqs. (7.33) and (7.36), the advantage of using Eq. (7.39) for evaluating the effect of non-normality on fatigue damage is that it is based on a easy closed-form formula for "by-hand" calculus.

7.5.2. Results based on Hermite model [Winterstein 1985 and 1988]

When we are restricted to mild non-linearities, we can use a first-order Hermite model, which gives the following expression of the non-normal damage intensity [Winterstein 1985]:

$$\overline{D}_{NB}^{nG} = \frac{v_0}{C} \left(\sqrt{2\sigma_X^2}\right)^k \Gamma\left(1 + \frac{k}{2}\right)\left[1 + k\,(k-1)\,\tilde{h}_4\right] \qquad (7.40)$$

in which $\tilde{h}_4 = (\gamma_4 - 3)/24$ is the coefficient appearing in Eq. (7.22); the first term in Eq. (7.40) is the damage intensity under Gaussian hypothesis, hence the remaining term in square brackets is the correction factor η accounting for the effect of non-normality:

$$\eta = \frac{k\,(k-1)\,(\gamma_4 - 3)}{24} \tag{7.41}$$

in which γ_4 is the value of kurtosis and k the slope of the S-N curve. As can be seen, in this model the skewness has no effect on damage.

Due to its first-order nature, errors in Eq. (7.41) may become more significant for larger values of k, or greater non-linearity. A second-order result was obtained, in which an approximate Weibull distribution is used for the range r in the non-Gaussian process. This leads to the following damage correction factor:

$$\eta = \left(\frac{\sqrt{\pi}\,K}{\Gamma(1+V_r)} \right)^k \cdot \frac{\Gamma(1+kV_r)}{\Gamma(1+k/2)} \tag{7.42}$$

where V_r (i.e. the coefficient of variation of the approximate Weibull distribution) is determined from the second-order softening Hermite model as:

$$V_r^2 = \frac{\sigma_r^2}{\mu_r^2} = \frac{4}{\pi}(1+h_4+\tilde{h}_4) - 1 \tag{7.43}$$

where h_4 and $\tilde{h}_4$ are given in Eqs. (7.23) and (7.24).

In the case $Z(t)$ is Gaussian, $K = 1$ and $h_4 = \tilde{h}_4 = 0$, so that Eq. (7.43) gives $V_r = 0.523$, consistent with the Rayleigh distribution (remind that for a standard Rayleigh distribution, the mean is $\mu = \sqrt{\pi/2}$ and the variance is $\sigma^2 = 0.429$, so that the coefficient of variation results as $\sigma/\mu = \sqrt{2 \cdot 0.429/\pi} = 0.5226$); Eq. (7.40) slightly overestimates the exact value $\eta = 1$ in this case, due to the approximate fit of the Weibull model).

Compared to the first-order model, the second order estimate of η in Eq. (7.43) is shown to accurately predict fatigue for larger values of k and γ_4.

In Figure 7.8 we show the effect of non-normality on the damage intensity for a narrow-band process for different values of the slope k of the S-N curve, as predicted by the Hermite and power-law models.

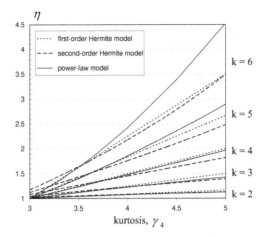

Figure 7.8: Effect of kurtosis γ_4 on the damage intensity for a narrow-band random process, for different values of the slope k of the S-N curve.

Chapter 8

FATIGUE ANALYSIS OF REAL COMPONENTS

8.1. INTRODUCTION

The wide literature concerning the fatigue analysis of random loadings by spectral methods is generally restricted to numerical simulations, for both Gaussian and non-Gaussian case studies [Wirsching and Light 1980, Bishop and Sherrat 1990, Bouyssy et al. 1993, Petrucci and Zuccarello 1999, Tovo 2002].

The well-known spectral representation approach (based on the Fast Fourier Transform technique) is used for simulating time histories that are stationary, asymptotically Gaussian and ergodic[*] [Shinozuka and Jan 1972, Shinozuka and Deodatis 1991, Grigoriu 1993, Hu and Schiehlen 1997].

As said in Chapter 7, the Gaussian assumption can be a strong limitation in real engineering case studies, where data are often non-Gaussian.

The simulation of non-Gaussian sample loadings, for example, can be performed by introducing in the spectral-based simulation methodology a time-invariant (memoryless) transformation technique (as for example Hermite models) [Gurley et al. 1996, Deodatis et al. 2001, Deodatis and Micaletti 2001].

In any case, both simulation approaches based on spectral representation generate random loadings that are stationary and this aspect can be of some advantage, since spectral methods for fatigue life assessment only apply to strictly stationary random processes.

On the other hand, such a class of stationary loadings is rarely encountered in real practical applications, since the loading responses measured on real components are generally not perfectly stationary (e.g. short transients or impact loadings may be present) due to the time-variability of the environmental conditions.

Consequently, two problems arise: firstly, the need of appropriate simulation techniques able to also reproduce non-stationary events (wavelet-based technique may be of some help [Gurley and Kareem 1999]), secondly, the problem of validation of spectral

---

[*] In WAFO, the routine `spec2sdat` may be used (which also implements the method described in [Dietrich and Newsam 1997]).

methods with random loadings measured on real structures or components, where the hypothesis of stationarity is not perfectly matched.

This Chapter shows the application of the analysis to the case of non-Gaussian random loadings (as described in Chapter 7), in order to validate the capability of including the load non-normality into the estimated rainflow cycle distribution. At the same time, this study aims to evaluate the robustness and the accuracy of the methods when applied to time histories measured on real components.

Two types of data will be considered: load time histories measured on a Mountain-bike in off-road tracks and load data taken from an automotive application.

The entire procedure of analysis may be divided in two phases. In the first one, the most important characteristics of the signal are considered and analysed by a time-domain (level-crossing spectrum, time-varying variance analysis) and time-frequency domain procedure (time-frequency spectrum, spectral density). The aim is to identify non-stationarities and then to extract sub-segments that more closely support the stationarity hypothesis.

Once stationary (or almost stationary) time histories have been identified, they will be analysed to evaluate the distribution of rainflow cycles and fatigue damage (under the linear rule), by including (if necessary) the deviation from the Gaussian behaviour.

8.2. GENERAL PROCEDURE OF ANALYSIS

Let $\tilde{z}(t)$ be the internal loading acting in a mechanical component, as obtained by measurements (e.g. during the service life) or simulations. Generally, it is assumed that $\tilde{z}(t)$ is a non-stationary and non-Gaussian loading.

The non-stationary nature of the internal responses can be produced by a modification of the external input loadings acting on the component, induced by a variation in the environmental conditions, e.g. a vehicle moving at different speeds or on different road types, different wind velocities during the day, etc. [Johannesson et al. 1995, Rouillard and Sek 2000, Rouillard 2002, Abdullah et al. 2004].

It is not immediate to classify the type of non-stationarities that may be encountered in practice. A simple model can be:

$$\tilde{z}(t) = \mu(t) + a(t)\,u(t) \tag{8.1}$$

where $u(t)$ is a zero-mean stationary process and $\mu(t)$, $a(t)$ are random or non-random (i.e. deterministic) slowly varying functions of time modelling the changes in mean and variance (see for example [Johannesson et al. 1995]).

In more complex (and also realistic) situations, the deviation from stationarity is due to the presence of local events, e.g. impact inputs or local transients.

The non-Gaussian nature of the internal stresses, instead, may be caused by non-normal external input loadings (e.g. wave or wind input loads) [Ochi and Ahn 1994]. Otherwise, even with external Gaussian loads, a non-linear structural response may induce internal forces or stresses to be non-Gaussian.

The results presented in this Chapter confirm that in some circumstances the non-Gaussian nature of the internal loadings may considerably increase the fatigue damage accumulation rate.

8.2.1. Analysis of load non-stationarity

All Gaussian or non-Gaussian spectral methods for fatigue life assessment are applicable only to stationary processes, which have statistical properties that do not vary with time. According to the above discussion, a measured load $\breve{z}(t)$ may be a time history taken from a non-stationary (and also non-Gaussian) random process (see for example Figure 8.2).

The first step of the analysis should indicate to what extent is the load non-stationary and whether it is possible to extract stationary or almost-stationary sub-segments from the whole measurement.

Different signal processing techniques can be used to detect strong or mildly non-stationarities in a given measured loading.

A commonly used method is called the Short-Time Fourier Transform (STFT): a moving window moves through a signal and the Fourier transform, applied to data inside the window, gives the local frequency content of the signal. Different window overlapping is also possible. The main shortcoming is that high resolution cannot be obtained in both time- and frequency-domain simultaneously, because of the inverse relation between window length and the corresponding frequency bandwidth. An example of the time-varying spectrum obtained with the STFT is shown in Figure 8.3 and Figure 8.5 for the forces measured in two different locations on a Mountain-bike during an off-road track.

The load non-stationarity can also be identified by checking the time evolution of the variance or the irregularity factor IF of the load, through a moving window as described above. Figure 8.4 and Figure 8.6 show an example of this analysis.

8.2.2. Analysis of load non-Gaussianity

Once the above analysis is completed, a stationary (or almost stationary) time history $z(t)$ could be identified and extracted from the entire measurement $\breve{z}(t)$ (it may also happen that $z(t)$ coincides with $\breve{z}(t)$).

Our more general assumption is that $z(t)$ represents a time history taken from a stationary non-Gaussian loading $Z(t)$.

A simple way to identify the deviation from the normal distribution is to use a normal probability plot, which compares the physical variable with the Gaussian cumulative distribution function (plotted in double linear scale). Through this graphical "goodness-of-fit" method it is fairly easy to discriminate if a variable does follow or not the normal distribution. As an example, Figure 8.7(a) and Figure 8.8(a) qualitatively illustrate how the values of the forces measured on a Mountain-bike deviate from the normal distribution. The computation of the skewness and the kurtosis will finally quantify the magnitude of this deviation.

138

Another way to check load non-normality is the comparison of the level upcrossing spectrum (giving the number of upward crossings for each load level) with the analytical formula valid for Gaussian processes (i.e. Rice's formula). An example is presented in Figure 8.9, where a non-Gaussian load is shown to cross larger levels more times than expected in the Gaussian case.

If the loading is actually non-Gaussian, the influence of non-normality on cycle distribution and fatigue damage should be adequately taken into account, through the theoretical non-Gaussian fatigue analysis discussed in Chapter 7.

The fatigue analysis concentrates on the estimation of the rainflow cycle distribution, since the fatigue damage rate under the linear damage rule follows from the cycle distribution, see Chapter 3. The comparison of the rainflow cycle distribution is made in terms of the fatigue loading spectrum (normalised here to time unit):

$$ F_{RFC}(s) = v_p \int_s^{+\infty} p_a^{RFC}(x)\, dx \qquad (8.2) $$

where $p_a^{RFC}(s)$ is the marginal rainflow amplitude density, and v_p (the peak intensity) measures the intensity of counted cycles.

Furthermore, in all damage computations, we shall assume for the S-N curve $s^k N = C$ a slope $k = 3$ and $k = 5$, and a fatigue strength $C = 1$.

In the non-Gaussian analysis, the transformation adopted to link a measured non-Gaussian loading segment $z(t)$ to its underlying Gaussian one, $x(t)$ is chosen to be the Hermite model (see Chapter 7), since it accounts for both skewness and kurtosis values. Parameters of the transformation are estimated from the non-Gaussian loading $z(t)$.

For the Gaussian segment $x(t)$ we can estimate the Gaussian rainflow cumulative distribution, $H_{X,RFC}^{G}(x_p, x_v)$, by using a linear combination as in Eq. (7.3) of Chapter 7 which, as stated by Eq. (7.11), also represents the rainflow cumulative distribution, $H_{Z,RFC}^{nG}(z_p, z_v)$, of the original non-Gaussian load $z(t)$ (i.e. our final task). Bandwidth parameters used in computation of the b_{app} coefficient are evaluated from the spectral density $W_X(\omega)$, which is estimated from the Gaussian data $x(t)$.

The final result of the non-Gaussian analysis is represented by the distribution of rainflow counted cycles which includes the non-normality of the load, i.e. $H_{Z,RFC}^{nG}(z_p, z_v)$. Another possibility, instead, is to estimate the rainflow cycle distribution disregarding this non-normality, i.e. taking the load $z(t)$ as if it were Gaussian. This can be achieved by simply applying the method valid for Gaussian processes (see Chapter 4) to measured data $z(t)$: first estimate the spectral density $W_Z(\omega)$, evaluate bandwidth parameters and the approximate b_{app} coefficient, and then compute the approximate rainflow distribution as in Eq. (7.3). The Gaussian rainflow cycle distribution, $H_{Z,RFC}^{G}(z_p, z_v)$, is then obtained.

139

Differentiation of both the rainflow cumulative distributions, $H^{G}_{Z,RFC}(z_p, z_v)$ and $H^{nG}_{Z,RFC}(z_p, z_v)$, gives the corresponding Gaussian and non-Gaussian rainflow joint densities, $h^{G}_{Z,RFC}(z_p, z_v)$ and $h^{nG}_{Z,RFC}(z_p, z_v)$, respectively.

In conclusion, for the non-Gaussian loading segment $z(t)$, the theoretical analyses provide two cycle distributions: the $h^{nG}_{Z,RFC}(z_p, z_v)$ distribution which takes into account the non-normal character of the load, and the cycle distribution $h^{G}_{Z,RFC}(z_p, z_v)$ which treating the segment as it were Gaussian, disregards its non-normality. As we shall see later on, these two distributions could be quite different; for example, Figure 8.10 shows how the non-Gaussian density, $h^{nG}_{Z,RFC}(z_p, z_v)$, could retain much more probability towards higher cycles than the assumed Gaussian one, $h^{G}_{Z,RFC}(z_p, z_v)$.

From the estimated distributions $h^{nG}_{Z,RFC}(z_p, z_v)$ and $h^{G}_{Z,RFC}(z_p, z_v)$, two estimated rainflow loading spectra are computed: the non-Gaussian one, say $F^{nG}_{Z,RFC}(s)$, accounting for non-normalities, and the Gaussian one, say $F^{G}_{Z,RFC}(s)$, disregarding non-normalities of the load. These loading spectra are computed by inserting in Eq. (8.2) the marginal amplitude density, $p^{RFC}_{a}(s)$ is computed for both Gaussian and non-Gaussian distributions from distributions $h^{nG}_{Z,RFC}(z_p, z_v)$ and $h^{G}_{Z,RFC}(z_p, z_v)$, respectively.

<u>Remark</u>: in the definition of both estimated Gaussian and non-Gaussian loading spectra, we adopted the same expected rate of peaks v_p, as calculated for the Gaussian process $X(t)$, since that transformation $G(\cdot)$ preserves the number of peaks in the Gaussian load and in its transformed $Z(t) = G(X(t))$.

On the other hand, the classical time-domain deterministic rainflow analysis applied to measured non-Gaussian data $z(t)$ gives a set of counted cycles, from which a sample loading spectrum $\hat{F}_{Z,RFC}(s)$ can be computed.

Similarly, fatigue damage computation yields two values of the expected fatigue damage rate: $\overline{D}^{nG}_{Z,est}$ including non-Gaussian effects, and $\overline{D}^{G}_{Z,est}$ not including them. The deterministic analysis gives, instead, for the non-Gaussian loading $z(t)$ a damage value under the linear rule, which is indicated as $\hat{D}^{nG}_{Z}$. Similarly, the same analysis on the underlying Gaussian data $x(t)$ provides a damage value indicated as $\hat{D}^{G}_{X}$. The ratio of these two damage values (see Chapter 7) is:

$$\eta = \frac{\overline{D}^{\,nG}}{\overline{D}^{\,G}} \tag{8.3}$$

and it is used for quantifying the influence of non-normal effects on damage.

The steps of all analyses (the theoretical Gaussian and non-Gaussian and the deterministic rainflow one) may be summarised in the subsequent scheme:

T1) Gaussian analysis (treat $Z(t)$ as if it were Gaussian):

1g) estimate spectral density $W_Z(\omega)$ of $Z(t)$;

2g) compute α_1, α_2 spectral parameters and b_{app} coefficient, Eq. (4.48);

3g) estimate rainflow Gaussian distribution $H^G_{Z,RFC}(z_p, z_v)$ for $Z(t)$, using a linear combination as in Eq. (7.3);

4g) differentiate to get the rainflow Gaussian density $h^G_{Z,RFC}(z_p, z_v)$;

5g) compute damage intensity $\overline{D}^G_{Z,est}$ (Gaussian).

T2) non-Gaussian analysis (treat $Z(t)$ as if it were non-Gaussian):

1ng) estimate skewness γ_3 and kurtosis γ_4 of non-Gaussian data $z(t)$;

2ng) estimate the inverse transformation $g(\cdot)$ and transform back to underlying Gaussian data $x(t)$;

3ng) estimate the spectral density $W_X(\omega)$ from $x(t)$ data; compute α_1 and α_2 bandwidth parameters and b_{app} coefficient, Eq. (4.48);

4ng) estimate as in Eq. (7.3) the rainflow cumulative distribution for the Gaussian data $H^G_{X,RFC}(x_p, x_v)$, which also represents the rainflow distribution for the transformed non-Gaussian data $z(t)$, i.e.

$$H^{nG}_{Z,RFC}(z_p, z_v) = H^{nG}_{Z,RFC}\big(G(x_p), G(x_v)\big) = H^G_{X,RFC}(x_p, x_v);$$

5ng) differentiate to get the non-Gaussian rainflow density $h^{nG}_{Z,RFC}(z_p, z_v)$;

6ng) compute damage $\overline{D}^{nG}_{Z,est}$ (non-Gaussian).

D) Deterministic analysis (rainflow analysis on loading $z(t)$ or $x(t)$):

1r) apply the rainflow count to $z(t)$ (or $x(t)$);

2r) compute rainflow cumulative $\hat{F}_{Z,RFC}(s)$ (or $\hat{F}_{X,RFC}(s)$);

In the **T2** analysis, fatigue calculations include the non-normality of time history $z(t)$. Instead, in the **T1** analysis the same loading history $z(t)$ is treated as if it were Gaussian, i.e. we neglect its deviation from Gaussianity; spectral density $W_Z(\omega)$ used to determine the Gaussian rainflow cycle distribution is estimated from $z(t)$ data.

8.3. MOUNTAIN-BIKE DATA

The theoretical aspects presented so far are now used in the fatigue analysis of a real mechanical component. Field measurements have been conducted on a Mountain-bike in a typical off-road use. A commercial Mountain-bike, with an Al 5085 TIG-welded frame and a rigid front fork with a total weight of 12.2 kg, has been used for the tests. Measuring points were chosen as to monitor the main external loads acting on the bicycle frame and the handlebar (test conditions are described in [Petrone et al. 1996]).

The overall measured channel locations are given in Figure 8.1 with regard to the external loads, together with the positive directions assumed in the channel calibration [Petrone et al. 1996]. Particular care has been reserved to the loads applied to the bike front components. The handlebar loads have been recorded by simultaneously sampling the horizontal and vertical loads at the left and right grips.

Figure 8.1: Location of the recorded external load channels on the instrumented Mountain-bike.

Test tracks were typical of north Italian bike off-road circuits. Cycling conditions were selected as a mixture of typical off-road use: uphill, downhill, plane cycling, combined with different surface conditions (asphalt, gravel, cobblestone), in both seated and standing conditions. Figure 8.2 shows two loading histories $\tilde{z}(t)$ measured during the same test track. Table 8.1 lists the most important characteristics of the test tracks used in the present study [Benasciutti and Tovo 2005b].

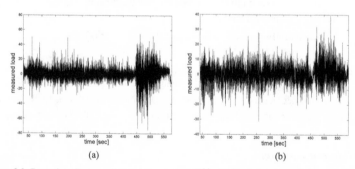

| (a) | (b) |

Figure 8.2: Examples of measured non-stationary load $\tilde{z}(t)$: (a) Track A – Force 1 (force on the fork); (b) Track A – Force 4 (vertical force on the handlebar).

From the description given above, it is expected that each load acquisition $\tilde{z}(t)$ represents a non-stationary load. Therefore, a preliminary analysis is needed to identify and eventually to extract a set of stationary (or almost stationary) time histories $z(t)$.

The analysis mainly concentrates on the time-varying spectrum (via STFT) and on the analysis of the load variance. From Figure 8.3 through Figure 8.6 we show the results of the time-varying analysis on two forces (i.e. the force on the fork and the verti-

142

cal force on the handlebar) measured on the Mountain-bike during the same test track (i.e. Track A). Window is 15 sec, with 90% overlapping.

Table 8.1: Characteristics of field measurements used in the present study.

Type of track	Time	Track	Type of surface	Cycling condition
Track A	0 – 100	PLN	Asphalt	Seated
	100 – 442	UPH	Gravel	Seated
	442 – 442	DWH	Cobblestone	Seated
	515 – 570	PLN	Cobblestone + asphalt	Seated
Track B	0 – 45	PLN	Asphalt	Standing
	45 – 95	PLN	Gravel	Seated
	95 – 460	UPH	Gravel	Seated
	460 – 548	DWH	Cobblestone	–
	548 – 570	PLN	Gravel	Seated
	570 – 600	PLN	Asphalt	Standing
Track C	0 – 94	DWH	Cobblestone	–
	94 – 230	UPH	Cobblestone	Various conditions
	230 – 330	DWH	Cobblestone	–

PLN = plane
UPH = uphill
DWH = downhill

A pronounced change in both the frequency content and the variance is detected in the force on the fork, corresponding to the uphill-downhill transition occurring at 100 sec (see description of Track A in Table 8.1). The same change is not so evident for the vertical handlebar force (only the spectrum looks quite similar), indicating that some differences may characterise the structural response to the same external input (the action on the bicycle fork probably reflects more directly the response to a change in the intensity of external forces).

The presented results also confirm that, in general, it is not so easy to detect single non-stationary events occurring in an observed measurement, when based only on the time-varying analysis.

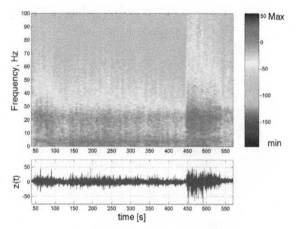

Figure 8.3: Time-frequency spectrum. Window: 15 sec, overlap: 90%; data from Track A – Force 1 (force on the fork).

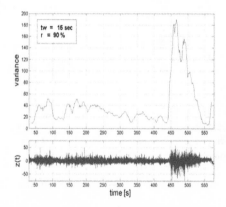

Figure 8.4: Time-varying variance. Window: 15 sec, overlap: 90%; data from Track A – Force 1 (force on the fork).

144

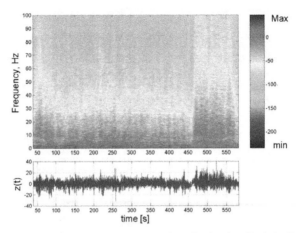

Figure 8.5: Time-frequency spectrum. Window: 15 sec, overlap: 90%; data from Track A – Force 4 (vertical force on the handlebar).

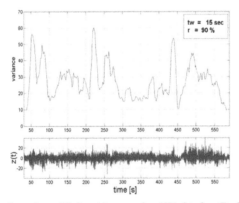

Figure 8.6: Time-varying variance. Window: 15 sec, overlap: 90%; data from Track A – Force 4 (vertical force on the handlebar).

In this particular application, a detailed report written during each test (see Table 8.1) was available. Based on this report, from each complete track measurement $\bar{z}(t)$ we extracted single segments $z(t)$, which were assumed as representative of stationary data. In Table 8.2 we summarise the characteristics of the complete dataset of stationary (or almost stationary) segments used in the present analysis.

The realisations $z(t)$ listed in Table 8.2 are taken from different stationary, yet non-Gaussian processes; this is confirmed for example by the normal probability plot shown in Figure 8.7(a), and also by the comparison between the observed and the theoretical Gaussian (i.e. Rice's formula) upcrossing spectrum (see Figure 8.9).

145

Table 8.2: Description of all stationary (or almost stationary) data used in the non-Gaussian analysis.

Code	Load History Code	Action	Type of track	Length [sec]
1	Track A – Force 1 – History 1	FK	PLN	80
2	Track A – Force 1 – History 2	FK	UPH	330
3	Track A – Force 2 – History 1	FR	PLN	80
4	Track A – Force 2 – History 2	FR	UPH	308
5	Track A – Force 3 – History 1	FR	DWH	75
6	Track A – Force 4 – History 1	HB – V	UPH	330
7	Track A – Force 5 – History 1	HB – V	PLN + UPH	410
8	Track A – Force 6 – History 1	HB – H	PLN	70
9	Track A – Force 6 – History 2	HB – H	UPH	340
10	Track B – Force 3 – History 1	FR	UPH	340
11	Track B – Force 3 – History 2	FR	UPH	375
12	Track B – Force 1 – History 1	FK	UPH	365
13	Track B – Force 1 – History 2	FK	PLN + UPH	430
14	Track B – Force 4 – History 1	HB – V	PLN + UPH	430
15	Track B – Force 4 – History 2	HB – V	DWH + PLN	140
16	Track C – Force 1 – History 1	FK	DWH	88
17	Track C – Force 1 – History 2	FK	UPH	116
18	Track C – Force 1 –History 3	FK	DWH	100

FK = Fork	Action:	PLN = plane	
FR = Frame	V = vertical	UPH = uphill	
HB = Handlebar	H = horizontal	DWH = downhill	

As can be seen, sometimes data can strongly deviate from the normal distribution, represented by the straight line in the normal probability paper and by the smooth line in the upcrossing spectrum.

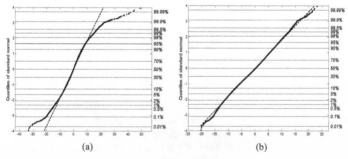

Figure 8.7: Normal probability papers: (a) measured non-Gaussian data $z(t)$ ($\gamma_3 = 0.19$, $\gamma_4 = 5.80$); (b) underlying Gaussian data $x(t)$. The straight line indicates the Gaussian distribution. Data refer to Track A – Force 1 – History 1 (force on the bicycle fork).

146

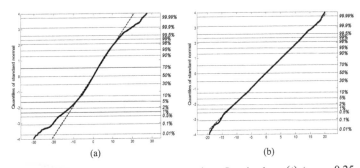

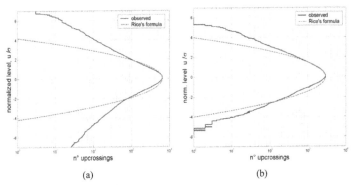

Figure 8.8: Normal probability papers: (a) measured non-Gaussian data $z(t)$ ($\gamma_3 = -0.25$, $\gamma_4 = 4.49$); (b) underlying Gaussian data $x(t)$. The straight line indicates the Gaussian distribution. Data refer to Track B – Force 4 – History 1 (handlebar vertical force).

Figure 8.9: Observed and Gaussian level upcrossing spectrum for $z(t)$. Data refer to: (a) track A – Force 1 – History 1 (force on the fork) ($\gamma_3 = 0.19$, $\gamma_4 = 5.80$); (b) track A – Force 4 – History 1 (handlebar vertical force) ($\gamma_3 = -0.25$, $\gamma_4 = 4.49$).

The computation of skewness and kurtosis values quantifies the magnitude of this deviation (see fourth and fifth column in Table 8.3) and confirms us that the complete set of stationary loading segments in Table 8.2 $z(t)$ must be treated as non-Gaussian.

The aim is now to estimate the rainflow cycle distribution for each time history $z(t)$ belonging to the set of non-Gaussian data listed in Table 8.3.

Each measured non-Gaussian data $z(t)$ is transformed to the underlying Gaussian data $x(t)$ by using the Hermite model described in Chapter 7. Parameters of the transformation are estimated for each non-Gaussian loading $z(t)$. Data reported in Table 8.3 show that the transformation preserves the variance, as well as the number of peaks and mean upcrossings, since the variance and the irregularity factor for both non-Gaussian and Gaussian processes are virtually the same.

After transforming we should theoretically obtain a set of Gaussian data $x(t)$, all having $\gamma_3 = 0$ and $\gamma_4 = 3$ (see also the normal probability paper in Figure 8.7(b) and Figure 8.8(b)). However, transformed data $x(t)$ are sometimes not perfectly Gaussian, as confirmed by the values of skewness and kurtosis listed in eighth and ninth column of Table 8.3. This fact is strictly related to the degree of non-normality of measured data $z(t)$ and to the level of efficiency of the transformation in correcting such non-linearities and could represent a source of inaccuracies in the subsequent cycle assessment procedure.

Table 8.3: Comparison of skewness γ_3 and kurtosis γ_4 for measured non-Gaussian data $z(t)$ and their related Gaussian $x(t)$ obtained via transformation $g(\cdot)$.

Code	non-Gaussian data, $z(t)$				Gaussian data, $x(t)$			
	IF	σ_Z^2	γ_3	γ_4	IF	σ_X^2	γ_3	γ_4
1	0.525	30.2	0.19	5.80	0.521	30.8	-0.07	2.84
2	0.293	27.3	0.41	4.86	0.293	27.8	-0.04	2.84
3	0.563	12.9	0.45	6.69	0.569	13.4	-0.04	2.71
4	0.334	11.3	-0.09	8.74	0.334	12.2	0.08	2.53
5	0.061	9.8	-0.15	3.84	0.062	9.8	0.00	3.04
6	0.135	27.4	-0.19	4.54	0.136	27.4	0.02	2.99
7	0.173	10.7	-0.01	5.48	0.173	10.7	0.01	2.97
8	0.240	15.5	0.44	6.78	0.252	15.2	0.09	3.13
9	0.176	15.7	0.28	4.48	0.180	15.6	0.00	3.05
10	0.274	10.0	0.01	7.93	0.275	10.7	0.13	2.55
11	0.274	10.3	-0.20	10.58	0.274	11.7	0.15	2.33
12	0.211	33.3	0.45	4.39	0.210	33.7	-0.01	2.82
13	0.224	34.2	0.42	4.29	0.224	34.6	-0.02	2.85
14	0.130	28.1	-0.25	4.49	0.132	28.1	0.00	3.01
15	0.219	27.7	0.46	4.49	0.223	27.7	0.01	3.03
16	0.599	75.4	0.05	4.65	0.599	74.7	-0.02	3.13
17	0.247	25.2	-0.02	5.02	0.247	25.0	-0.09	3.08
18	0.516	51.7	0.16	5.36	0.516	52.1	-0.01	2.93

Anyway, by comparing the skewness and kurtosis values in Table 8.3 calculated for measured data $z(t)$ and for their transformed $x(t)$, we can conclude that the Hermite model is sufficiently accurate in correcting mild and strong non-linearities presented by our dataset.

At this point, we can apply all steps of the analysis described at the beginning of this Chapter, in order to obtain two joint densities of rainflow cycles: the non-Gaussian, $h_{Z,\text{RFC}}^{nG}(z_p, z_v)$, and the Gaussian, $h_{Z,\text{RFC}}^{G}(z_p, z_v)$ one.

As can be seen in Figure 8.10, these two distributions may be quite different, with the non-Gaussian density retaining much more probability towards higher cycles than the Gaussian one. The corresponding cumulative spectra may be also computed: the non-Gaussian, $F_{Z,\mathrm{RFC}}^{\mathrm{nG}}(s)$, and the Gaussian one, $F_{Z,\mathrm{RFC}}^{\mathrm{G}}(s)$.

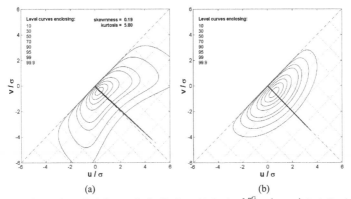

(a)   (b)

Figure 8.10: Comparison of rainflow cycle distributions: (a) density $h_{Z,\mathrm{RFC}}^{\mathrm{nG}}(z_{\mathrm{p}}, z_{\mathrm{v}})$ including loading non-normality and (b) density $h_{Z,\mathrm{RFC}}^{\mathrm{G}}(z_{\mathrm{p}}, z_{\mathrm{v}})$ disregarding it. Representation of the Dirac delta function is qualitative. Data refer to Track A – Force 1 – History 1.

As can be seen in Figure 8.11 and Figure 8.12, the estimated spectrum under the Gaussian hypothesis, $F_{Z,\mathrm{RFC}}^{\mathrm{G}}(s)$, is never coincident with the loading spectrum obtained from measured data, meaning that deviation from the Gaussian behaviour could reflect into more probability of occurrence of larger cycles than predicted under the Gaussian hypothesis (we note that all data are leptokurtic, $\gamma_4 > 3$).

On the contrary, the estimated non-Gaussian spectrum obtained from the TB method, $F_{Z,\mathrm{RFC}}^{\mathrm{nG}}(s)$, which accounts for non-normalities in the load, seems to agree better with experimental results: the shape of the estimated non-Gaussian spectra is closer to the experimental spectrum than the Gaussian one. This is particularly evident in extrapolating both Gaussian and non-Gaussian spectra towards higher amplitudes, where the Gaussian estimation gives an incorrect prediction of the rate of occurrence of the largest cycles.

149

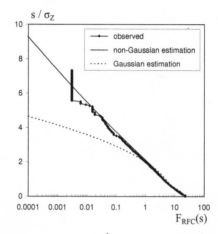

Figure 8.11: The observed sample spectrum $\hat{F}_{Z,\text{RFC}}(s)$ compared with the estimated Gaussian $F_{Z,\text{RFC}}^{G}(s)$ and non-Gaussian $F_{Z,\text{RFC}}^{nG}(s)$ loading spectra. Track A – Force 1 – History 2.

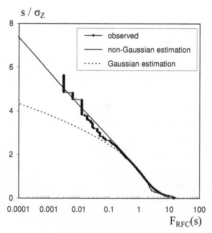

Figure 8.12: The observed sample spectrum $\hat{F}_{Z,\text{RFC}}(s)$ compared with the estimated Gaussian $F_{Z,\text{RFC}}^{G}(s)$ and non-Gaussian $F_{Z,\text{RFC}}^{nG}(s)$ loading spectra. Track A – Force 4 – History 1.

Finally, we compare the damage intensity calculated under the Palmgren-Miner damage law, by assuming or the S-N curve we take $C = 1$, and $k = 3$ and $k = 5$.

Table 8.4: Damage $\hat{D}_Z^{nG}$ is from measured non-Gaussian data $z(t)$; the theoretical damage estimations are $\overline{D}_{Z,\text{est}}^{nG}$ (non-Gaussian) and $\overline{D}_{Z,\text{est}}^{G}$ (Gaussian). Damage $\hat{D}_X^{G}$ is Gaussian data $x(t)$ (S-N slope: $k = 3$).

Code	$\hat{D}_Z^{nG}$	$\hat{D}_X^{G}$	η	$\overline{D}_{Z,\text{est}}^{nG}$	Var. %	$\overline{D}_{Z,\text{est}}^{G}$	Var. %
1	11067	7485	1.478	15239	37.7	9603	-13.2
2	4875	3538	1.378	5756	18.1	4170	-14.5
3	4074	2800	1.455	5872	44.2	3177	-22.0
4	2451	1366	1.794	4026	64.3	1706	-30.4
5	2487	2104	1.182	2612	5.0	2321	-6.7
6	1431	1154	1.240	1611	12.6	1199	-16.2
7	600	427	1.405	749	24.9	509	-15.2
8	2337	1333	1.754	2427	3.8	1651	-29.4
9	1025	789	1.298	1201	17.2	956	-6.7
10	1247	825	1.511	2125	70.5	982	-21.3
11	1687	916	1.842	3429	103.2	1050	-37.8
12	4085	3222	1.268	5288	29.5	4016	-1.7
13	4573	3628	1.260	5258	15.0	4254	-7.0
14	1526	1225	1.245	1829	19.9	1362	-10.8
15	3943	2989	1.319	4023	2.0	3423	-13.2
16	39789	30112	1.321	42930	7.9	33230	-16.5
17	3818	2656	1.438	3391	-11.2	2979	-22.0
18	23403	16056	1.458	28354	21.2	20098	-14.1

The deterministic rainflow analysis on measured data $z(t)$ gives a damage value which is indicated as $\hat{D}_Z^{nG}$. Damage values from the theoretical analysis are $\overline{D}_{Z,\text{est}}^{nG}$ (including non-Gaussian effects) and $\overline{D}_{Z,\text{est}}^{G}$ (not including them). Also a damage value, indicated as $\hat{D}_X^{G}$, given by the deterministic rainflow analysis on transformed (virtually Gaussian) $x(t)$ data, is inserted.

Comparison of the damage values (see second and third column of Table 8.4) confirm us that the damage intensity observed in a non-Gaussian process is generally greater than that calculated in a Gaussian process with the same variance; this aspect, due to the fact that all collected data are leptokurtic (with kurtosis $\gamma_4 > 3$), is well-established in literature (see for example [Winterstein 1985, Sarkani et al. 1994]). In particular, the damage ratio η defined in Chapter 7 and reported in Eq. (8.3) is always greater than one and tends to increase with the kurtosis value, since larger cycles are more damaging.

151

Table 8.5: Damage $\hat{D}_Z^{nG}$ is from measured non-Gaussian data $z(t)$; the theoretical damage estimations are $\overline{D}_{Z,est}^{nG}$ (non-Gaussian) and $\overline{D}_{Z,est}^{G}$ (Gaussian). Damage $\hat{D}_X^{G}$ is Gaussian data $x(t)$ (S-N slope: $k = 5$).

Code	$\hat{D}_Z^{nG}$	$\hat{D}_X^{G}$	η	$\overline{D}_{Z,est}^{nG}$	Var. %	$\overline{D}_{Z,est}^{G}$	Var. %
1	4417336	980032	4.507	6936917	57.0	1325445	-70.0
2	1360610	428157	3.178	1697816	24.8	520843	-61.7
3	706921	146305	4.832	1368673	93.6	176951	-75.0
4	664095	62208	10.675	1362279	105.1	86770	-86.9
5	160925	93569	1.720	162570	1.0	98942	-38.5
6	310277	131401	2.361	387427	24.9	150285	-51.6
7	68214	17942	3.802	98625	44.6	24946	-63.4
8	524623	92695	5.660	564654	7.6	108614	-79.3
9	134626	54280	2.480	166170	23.4	69456	-48.4
10	237595	29007	8.191	467197	96.6	41737	-82.4
11	681309	33595	20.280	1380274	102.6	46278	-93.2
12	1082676	467990	2.313	1735439	60.3	655504	-39.5
13	1200638	546735	2.196	1518679	26.5	665800	-44.5
14	337050	146544	2.300	469508	39.3	181444	-46.2
15	869667	372194	2.337	868531	-0.1	406535	-53.3
16	24662363	10321639	2.389	29718200	20.5	11111855	-54.9
17	991198	302837	3.273	674979	-31.9	300107	-69.7
18	14493924	3772720	3.842	18869342	30.2	4837392	-66.6

The damage non-Gaussian $\overline{D}_{Z,est}^{nG}$ seems to agree well with damage $\hat{D}_Z^{nG}$ from deterministic rainflow analysis, even if it sometimes slightly overestimates it. This can be attributed probably to the finite length of measured data, in which large cycles (which cause most of the damage), since are rarely present, don't contribute to the total experimental damage. On the contrary, the Gaussian damage $\overline{D}_{Z,est}^{G}$ generally underestimates deterministic damage $\hat{D}_Z^{nG}$. This conclusion is valid for both slopes k, even if differences are more pronounced when $k = 5$, see Table 8.5, due to higher contribution on damage given by larger cycles.

8.4. AUTOMOTIVE APPLICATION

An automotive engineering application is considered, in which the loading response of a structural detail under realistic loading conditions is studied by numerical simulations [Bel Knani et al. 2007].

The local biaxial state of stress relative to five points into the automotive component is considered. The stress tensor components are $\{\sigma_{xx}, \sigma_{yy}, \tau_{xy}\}$, where σ_{xx} and σ_{yy} denote tensile stresses, while τ_{xy} is a shear stress. For each point, a stress tensor time history is simulated and conveniently normalised, so that all components have a zero mean value, and so that the maximum variance amongst them is equal to unity (see Table 8.6).

Since this application concerns a complex (i.e. biaxial) state of stress, the first step of the analysis is the definition of a proper uniaxial stress quantity, on which to perform all fatigue calculations. As a first approximation, this quantity is:

- the most significant stress amongst σ_{xx}, σ_{yy} and τ_{xy} stress components (e.g. stress σ_{yy} in Figure 8.13(a));

- an equivalent shear stress, say τ_{eq}, determined as a function of the stress tensor components $\{\sigma_{xx}, \sigma_{yy}, \tau_{xy}\}$ (as in the case of Figure 8.13(b)).

This uniaxial quantity will be generically denoted as $z(t)$.

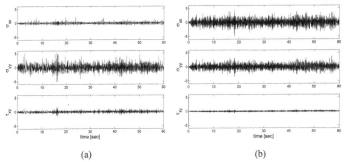

(a)   (b)

Figure 8.13: Stress components σ_{xx}, σ_{yy} and τ_{xy}. (a) data D1; (b) data E1.

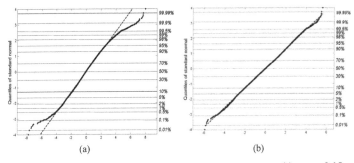

(a)   (b)

Figure 8.14: Normal probability papers: (a) Measured non-Gaussian data $z(t)$ ($\gamma_3 = 0.18$, $\gamma_4 = 4.18$); (b) Underlying Gaussian data $x(t)$. The straight line indicates the Gaussian distribution. Data refer to D1-σ_{yy}.

153

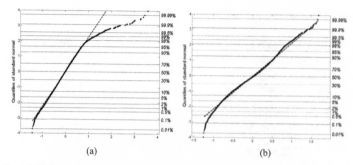

(a)   (b)

Figure 8.15: Normal probability papers: (a) Measured non-Gaussian data $z(t)$ ($\gamma_3 = 0.70$, $\gamma_4 = 6.13$); (b) Underlying Gaussian data $x(t)$. The straight line indicates the Gaussian distribution. Data refer to D1-σ_{xx}.

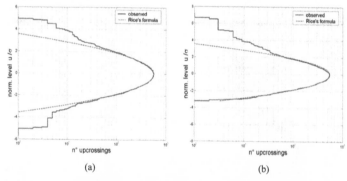

(a)   (b)

Figure 8.16: Observed and Gaussian level upcrossing spectrum for $z(t)$. Data refer to (a) D1-σ_{yy} ($\gamma_3 = 0.18$, $\gamma_4 = 4.18$); (b) D1-σ_{xx} ($\gamma_3 = 0.70$, $\gamma_4 = 6.13$).

The stress components generally deviate from the normal distribution, see for example the normal probability plot in Figure 8.14(a) and Figure 8.15(a)), and the level upcrossing spectrum in Figure 8.16. Computation of the skewness and kurtosis for stress component each time history confirms this observation, see Table 8.6 for stress components σ_{xx}, σ_{yy} and τ_{xy}, and Table 8.7 for the equivalent shear stress τ_{eq}.

As usual, we compare the cumulative rainflow spectrum $\hat{F}_{Z,RFC}(s)$ resulting from rainflow deterministic analysis on $z(t)$ data (i.e. rainflow count and Palmgren-Miner rule) with theoretical spectra $F_{Z,RFC}^{G}(s)$ and $F_{Z,RFC}^{nG}(s)$ provided by the Gaussian and non-Gaussian analysis, respectively.

Table 8.6: Characteristics of CRF data used in the analysis.

Code	σ_{xx}			σ_{yy}			τ_{xy}		
	σ^2	γ_3	γ_4	σ^2	γ_3	γ_4	σ^2	γ_3	γ_4
A1	0.12	-0.51	3.02	0.10	-0.39	2.80	1.00	0.42	2.86
A2	0.57	-0.22	2.91	1.00	-0.22	2.92	0.88	0.52	2.81
B1	0.02	-0.02	3.38	1.00	-0.29	2.88	0.02	0.06	3.27
B2	1.00	0.15	3.04	0.99	0.26	2.90	0.03	-0.28	3.17
C1	1.00	0.15	2.96	0.12	-0.16	3.13	0.01	0.09	3.82
C2	0.56	-0.15	2.95	1.00	-0.16	3.04	0.01	-0.13	4.01
D1	0.08	0.70	6.13	1.00	0.18	4.18	0.07	0.18	4.71
D2	0.06	-0.13	3.78	1.00	0.06	3.58	0.25	0.27	3.90
E1	1.00	0.04	3.41	0.65	0.11	3.38	0.02	0.11	3.52
E2	1.00	-0.03	3.44	0.24	0.07	4.73	0.06	-0.06	3.73

Table 8.7: Characteristics of the equivalent shear stress τ_{eq} resulting from CRF data.

Code	τ_{eq}		
	σ^2	γ_3	γ_4
A1	1.00	-0.42	2.86
A2	0.90	0.51	2.81
B1	0.28	0.21	2.94
B2	0.03	0.27	3.19
C1	0.46	0.15	3.00
C2	0.02	0.15	3.29
D1	0.33	-0.09	4.19
D2	0.53	-0.01	3.47
E1	0.05	0.08	4.22
E2	0.39	-0.10	3.41

The non-Gaussian stress time history $z(t)$ is transformed to the underlying Gaussian history $x(t)$ through the Hermite model. Figure 8.14(b) Figure 8.15(b) confirm that $x(t)$ is almost Gaussian.

Figure 8.17 and Figure 8.18 show the comparison amongst rainflow fatigue spectra, proving how the non-Gaussian analysis is more accurate in evaluating the rainflow cycle distribution in non-Gaussian random loadings.

Finally, fatigue damage is computed for some data, by assuming a S-N curve with fatigue strength $C = 1$; the results reported in Table 8.8 refer to a slope $k = 3$, whereas those reported in Table 8.9 refer to a slope $k = 5$.

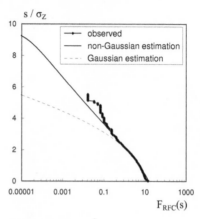

Figure 8.17: The observed sample spectrum $\hat{F}_{Z,\text{RFC}}(s)$ compared with the estimated Gaussian $F_{Z,\text{RFC}}^{G}(s)$ and non-Gaussian $F_{Z,\text{RFC}}^{nG}(s)$ loading spectra. Data are D1-σ_{yy}.

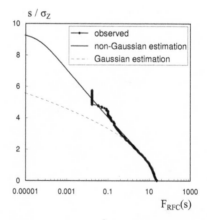

Figure 8.18: The observed sample spectrum $\hat{F}_{Z,\text{RFC}}(s)$ compared with the estimated Gaussian $F_{Z,\text{RFC}}^{G}(s)$ and non-Gaussian $F_{Z,\text{RFC}}^{nG}(s)$ loading spectra. E1-τ_{eq}.

As can be seen, the non-Gaussian damage $\overline{D}_{Z,\text{est}}^{\text{nG}}$ is generally closer to the deterministic damage $\hat{D}_{Z}^{\text{nG}}$, than the Gaussian damage $\overline{D}_{Z,\text{est}}^{G}$. In particular, the difference is sensitive to the slope k.

Table 8.8: Damage $\hat{D}_{Z}^{\text{nG}}$ is from measured non-Gaussian data $z(t)$; the theoretical damage estimations are $\overline{D}_{Z,\text{est}}^{\text{nG}}$ (non-Gaussian) and $\overline{D}_{Z,\text{est}}^{G}$ (Gaussian). Damage $\hat{D}_{X}^{G}$ is Gaussian data $x(t)$ (S-N slope: $k = 3$).

Code	Stress	$\hat{D}_{Z}^{\text{nG}}$	$\hat{D}_{X}^{G}$	η	$\overline{D}_{Z,\text{est}}^{\text{nG}}$	Var. %	$\overline{D}_{Z,\text{est}}^{G}$	Var. %
D-1	σ_{yy}	45.1	34.3	1.318	42.4	-6	35.8	-21
D-2	τ_{xy}	34.6	28.5	1.215	37.6	9	32.5	-6
E-1	τ_{xy}	60.3	54.2	1.112	63.7	6	70.8	17
E-2	σ_{xx}	51.8	47.3	1.096	55.0	6	51.5	0
D-1	τ_{eq}	44.4	33.7	1.320	41.8	-6	35.9	-19
E-1	τ_{eq}	76.5	60.8	1.259	84.4	10	77.9	2

Table 8.9: Damage $\hat{D}_{Z}^{\text{nG}}$ is from measured non-Gaussian data $z(t)$; the theoretical damage estimations are $\overline{D}_{Z,\text{est}}^{\text{nG}}$ (non-Gaussian) and $\overline{D}_{Z,\text{est}}^{G}$ (Gaussian). Damage $\hat{D}_{X}^{G}$ is Gaussian data $x(t)$ (S-N slope: $k = 5$).

Code	Stress	$\hat{D}_{Z}^{\text{nG}}$	$\hat{D}_{X}^{G}$	η	$\overline{D}_{Z,\text{est}}^{\text{nG}}$	Var. %	$\overline{D}_{Z,\text{est}}^{G}$	Var. %
D-1	σ_{yy}	439.5	173.6	2.532	372.4	-15	191.8	-56
D-2	τ_{xy}	252.8	131.6	1.920	287.4	14	158.8	-37
E-1	τ_{xy}	367.1	252.5	1.454	390.6	6	378.9	3
E-2	σ_{xx}	306.1	223.6	1.369	332.3	9	250.4	-18
D-1	τ_{eq}	425.4	171.0	2.487	366.8	-14	192.6	-55
E-1	τ_{eq}	655.1	280.4	2.337	722.8	10	396.1	-40

Chapter 9

FATIGUE LIFE DATA: THEORETICAL ESTIMATIONS VS. EXPERIMENTAL RESULTS

9.1. INTRODUCTION

In the preceding chapters, we studied the main properties of the statistical distribution of cycles counted by the rainflow method and the fatigue damage under the linear damage rule.

Firstly, we focused on Gaussian random processes: several approaches were reviewed and numerical simulations were used to identify the most accurate ones (e.g. Dirlik's formula and TB method).

At the same time, we were conscious of the limitation imposed by the Gaussian hypothesis when dealing with real applications, since non-linearities present in real systems often cause internal stresses to be non-Gaussian. In fact, when trying to apply such methods to time histories taken from experimental measurements we noted that the Gaussian assumption was not verified, resulting in wrong estimates of the cycle distribution and corresponding not conservative predictions of the related fatigue damage.

The conclusion was that the non-Gaussian characteristics of real loadings must be included into the fatigue analysis. In Chapter 7, a further theoretical development of the TB method was proposed, which was able to include the non-Gaussian characteristics of the random loading into the fatigue assessment procedure. The accuracy of this type of non-Gaussian analysis was verified by considering data observed in real component in service, as for example the forces measured in a Mountain-bike on off-road tracks (see Chapter 8).

The last step to investigate is the accuracy of the estimation obtained by spectral methods when compared with results given by experimental tests. We can expect that results from experimental tests on specimens or real components should provide the most complete information concerning the real response of the material when subjected to random loadings, e.g. the non-linear damage accumulation process due to load interaction effects, which is not included in linear damage accumulation models.

This Chapter aims to compare the predictions obtained from some analytical computations with results obtained from experimental tests, concerning Gaussian and non-
Gaussian random loadings.

In general it is very difficult to collect experimental results concerning random loadings, since such kind of tests are generally very lengthy and costly. In addition, only a
part of data available in literature allows one to effectively reconstruct input parameters
of the random loading used in experiments (e.g. its spectral density and its probability
density function), in order to perform new numerical simulations and to compare them
with experimental results.

An interesting set of experimental data is presented in [Sarkani et al. 1994, Kihl et al.
1995, Sarkani et al. 1996], with other data given in [Colombi and Doliński 2001]: a
structural welded joint is tested under Gaussian and non-Gaussian random loading having a broad-band and a bimodal spectral density. The experimental lifetime is registered
as the number of cycles counted at failure.

In the following sections, we will report the main details of the experimental tests
performed in the cited references. Then, we will illustrate the results of the comparison
amongst the experimental tests, the results obtained in new simulations based on loadings used in experiments and the lifetime theoretical predictions given by some spectral
methods.

9.2. THE EXPERIMENTAL INVESTIGATIONS

In order to determine the effect of non-normality and high-frequency content of the
loading process on the rate of fatigue damage accumulation, an experimental investigation was carried out in [Kihl et al. 1995, Sarkani et al. 1996]. No attempts were performed to investigate the spectral density shape effect on the fatigue lifetime.

9.2.1. The tested specimen

The specimen used in experimental tests was the welded cruciform joint shown in
Figure 9.1. The vertical leg of the cruciform was 356 mm (14 in) long and the two horizontal stems were each 51 mm (2 in) long and were attached to the mid section by full
penetration gas metal arc weld fillet welds. The joints were 102 mm (4 in) wide.

The reported yield stress and ultimate stress of the rolled steel plate from which the
joints were made are 638 MPa (92.5 ksi) and 683 MPa (99.0 ksi)[*].

In the fatigue tests, the joint was loaded axially in the vertical direction with random
loads applied to the ends of the vertical legs by means of hydraulic grips.

In order to analyse the results of variable amplitude tests properly and to set analytical predictions, it was necessary to know the constant S-N curve for the joint. Classical
constant amplitude tests were conducted at four stress levels, the highest and the lowest
load levels produce nominal axial stresses of 310 MPa (45 ksi) and 83 MPa (12 ksi), respectively [Kihl et al. 1995].

---

[*] 1 ksi = 1000 lbf in^{-2} = 6.895 MPa

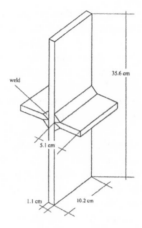

Figure 9.1: The cruciform joint used in experimental tests. (Reprinted from [Colombi and Doliński 2001], with permission from Elsevier).

Analysis of results gave a single-line S-N curve according to the expression:

$$s^{3.210} \, N \; = \; 10^{9.559} \qquad (\text{ksi units})$$

(9.1)

$$s^{3.210} \, N \; = \; 1.7811 \cdot 10^{12} \qquad (\text{MPa units})$$

where N is the number of cycles to failure and s is the stress amplitude (half a range), in MPa. Figure 9.2 shows the plot of the S-N curve (in ksi units) together with the constant amplitude test results.

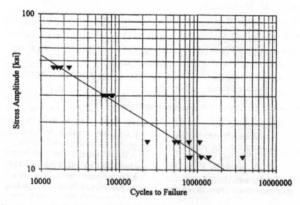

Figure 9.2: Constant amplitude S-N curve (in ksi units). (Reprinted from [Sarkani et al. 1996], with permission from Elsevier).

The S-N curve given above was used in conjunction with the rainflow algorithm to obtain variable amplitude fatigue life predictions for the loadings used in the experimental investigations.

9.2.2. The Gaussian and the non-Gaussian random loading

As described in [Kihl et al. 1995, Sarkani et al. 1996], the variable amplitude experimental tests used zero-mean Gaussian and non-Gaussian random loadings, which will be denoted (as in the notation of Chapter 7) as $X(t)$ and $Z(t)$, respectively.

The Gaussian loading $X(t)$ was generated by a standard frequency-domain simulation technique from a spectral density function. Two spectral densities were used, i.e. the broad-band spectrum:

$$W_X(\omega) = A \frac{12}{\omega^7} e^{-\frac{2.5}{\omega^3}} \qquad 0.75 \le \omega \le 3.0 \qquad (9.2)$$

and the bimodal spectrum:

$$W_X(\omega) = \begin{cases} A \dfrac{12}{\omega^7} e^{-\frac{2.5}{\omega^3}} & 0.75 \le \omega \le 3.0 \\[4mm] A \dfrac{1.2}{(\omega-6)^7} e^{-\frac{2.5}{(\omega-6)^3}} & 6.75 \le \omega \le 8.0 \end{cases} \qquad (9.3)$$

where A is a scaling factor used to set a chosen value for the loading variance σ_X^2 (i.e. the moment λ_0). These two spectra are plotted in Figure 9.3, by taking $\lambda_0 = 1$. Their main frequency-domain characteristics are reported in Table 9.1 and reveal how the bimodal spectrum is the most wide-banded spectrum (the lowest α_1 and α_2 values). Coefficient b_{app} for the TB method is defined in Chapter 4.

Table 9.1: Frequency-domain characteristics of spectral densities used in simulations: v_0 (mean upcrossing rate), v_p (peak frequency), α_1 and α_2 bandwidth parameters, b_{app} index.

Spectral density	v_0	v_p	α_1	α_2	b_{app}
broad-band	0.202	0.240	0.963	0.844	0.770
bimodal	0.390	1.004	0.713	0.389	0.666

In order to investigate the effect of the load intensity on the resulting fatigue lifetime, the Gaussian loading used in experiments had different values of the root mean square,

RMS, (i.e. the standard deviation σ_X). Three levels were investigated: 51.7 MPa (7.5 ksi), 69 MPa (10 ksi) and 103.4 MPa (15 ksi).

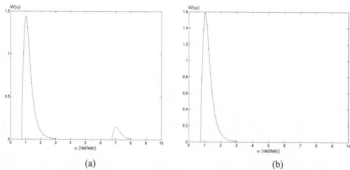

(a)   (b)

Figure 9.3: Spectral densities used in simulations ($\lambda_0 = 1$): (a) bimodal and (b) broad-band.

The non-Gaussian loading $Z(t)$ was obtained by transforming each extreme (i.e. maximum and minimum) of the basic simulated Gaussian loading $X(t)$, by the power-law model, described in Chapter 7:

$$Z = G(X) = \frac{X + \beta \left(\text{sgn}(X)\right)\left(\left| X \right|^n\right)}{D} \qquad (9.4)$$

where β and n are parameters controlling the degree of non-normality and D is introduced to force the condition $\sigma_X = \sigma_Z$. The transformation also preserves the number of peaks and of mean upcrossings, so assuring that $X(t)$ and $Z(t)$ will have the same irregularity factor, IF.

This model uses the kurtosis as a unique measure of non-normality. The parameters of the transformation are uniquely determined by specifying the kurtosis γ_4, the standard deviation and either n or β (see Chapter 7 for more details).

Two kurtosis values were investigated in experiments, namely $\gamma_4 = 2$ and $\gamma_4 = 5$. For the case $\sigma_X = \sigma_Z = 1$, the parameters are given in Table 9.2.

Table 9.2: Parameter values of the power-law model for two values of the kurtosis γ_4 for the case $\sigma_X = \sigma_Z = 1$ (table is reported from Chapter 7).

γ_4	n	β	D
2	0.5	1.735	2.527
5	2	0.342	1.563

The above parameters are only valid when $\sigma_X = \sigma_Z = 1$ and different values have to be used for other RMS values.

In our numerical simulations, we found more convenient to maintain above parameters and to scale the RMS value. To this scope, we introduce two auxiliary zero-mean random loadings, a Gaussian $\widetilde{X}(t)$ and a non-Gaussian $\widetilde{Z}(t)$ one, both having unit RMS (i.e. $\sigma_{\widetilde{X}} = \sigma_{\widetilde{Z}} = 1$). Process $\widetilde{X}(t)$ is obtained by scaling $X(t)$ to unit variance, whereas loading $\widetilde{Z}(t)$ is the transformed of $\widetilde{X}(t)$ as in Eq. (9.4), and has a kurtosis, say $\widetilde{\gamma}_4$, equal to 2 or 5. For a given RMS value, say $\overline{\sigma}$, proper scaling is obtained as:

$$X(t) = \overline{\sigma}\, \widetilde{X}(t)$$
$$Z(t) = \overline{\sigma}\, \widetilde{Z}(t)$$
(9.5)

Since for $Z(t)$ the following relations are valid:

$$\mu_Z = E[Z(t)] = \overline{\sigma}\, E[\widetilde{Z}(t)] = 0$$

$$\sigma_Z^2 = E[Z^2(t)] = \overline{\sigma}^2\, E[\widetilde{Z}^2(t)] = \overline{\sigma}^2$$
(9.6)

$$\gamma_4 = \frac{E[(Z(t) - \mu_Z)^4]}{\sigma_Z^4} = \frac{E[Z(t)^4]}{\sigma_Z^4} = \frac{E[\widetilde{Z}(t)^4]}{\overline{\sigma}^4} = \widetilde{\gamma}_4$$

we can see that $Z(t)$ has the chosen RMS value $\overline{\sigma}$ and also the same kurtosis $\widetilde{\gamma}_4$ of $\widetilde{Z}(t)$, as requested.

9.2.3. Experimental results

Numerical simulations were used in order to generate Gaussian and non-Gaussian loadings, $x(t)$ and $z(t)$, both having the same RMS value, and with the non-Gaussian loading having the desired kurtosis value ($\gamma_4 = 2$ or $\gamma_4 = 5$).

These loadings were repeated as many times as necessary to fail the specimen previously described. The length of the simulated block was chosen such that the desired statistics (moments) of the extremes matched the theoretical values.

Failure was defined to be when the current stiffness of the specimen was one half of the original one. When the stiffness reaches this value, cracks were growing so rapidly that total fracture of the specimen was imminent.

Each test was repeated up to 4 times and the geometric mean was evaluated. Due to the small size of the ensemble, no reliable second moment information (e.g. the sample variance) was available. Tables below gather all experimental results.

163

Table 9.3 reports test results under Gaussian loadings $x(t)$ (kurtosis = 3), whereas Table 9.4 and Table 9.5 list results under non-Gaussian loading $z(t)$ with kurtosis equal to 2 and 5, respectively.

Table 9.3: Gaussian fatigue test results (kurtosis = 3).

RMS MPa (ksi)	Cycles to failure				Geometric mean
	Individual experiments				
	Exp. 1	Exp. 2	Exp. 3	Exp. 4	
broad-band					
51.7 (7.5)	1202100	891100	913000	4047100	1410491
69 (10)	873800	420000	450800	425600	515124
103.4 (15)	185200	218300	202800	154700	188718
bimodal					
51.7 (7.5)	5235600	3019700	2992100	5057300	3932840
69 (10)	2415000	1625400	1787000	2410100	2027726
103.4 (15)	406000	436600	393300	320900	386746

Data in Table 9.4 are taken from [Colombi and Doliński 2001], which however reports only the geometric mean (the results for each experiment are not available).

Examination of the geometric means of the experimental results indicates that the fatigue lives generally are shorter with higher RMS values, for both the broad-band and the bimodal spectral density, as well as for all values of kurtosis. This fact can be also verified from experimental data presented in [Łagoda et al. 2001, Moreno et al. 2003].

Table 9.4: Non-Gaussian fatigue test results (kurtosis = 2).

RMS MPa (ksi)	Cycles to failure				Geometric mean
	Individual experiments				
	Exp. 1	Exp. 2	Exp. 3	Exp. 4	
broad-band					
51.7 (7.5)	–	–	–	–	2431900
69 (10)	–	–	–	–	888000
103.4 (15)	–	–	–	–	325350
bimodal					
51.7 (7.5)	–	–	–	–	6780100
69 (10)	–	–	–	–	3496000
103.4 (15)	–	–	–	–	666700

Table 9.5: Non-Gaussian fatigue test results (kurtosis = 5).

RMS MPa (ksi)	Cycles to failure				Geometric mean
	Individual experiments				
	Exp. 1	Exp. 2	Exp. 3	Exp. 4	
broad-band					
51.7 (7.5)	951800	742900	1067900	703000	853573
69 (10)	373800	326300	273000	301000	316407
103.4 (15)	47900	45100	39500	44600	44168
bimodal					
51.7 (7.5)	3376500	2218700	2941300	2107600	2610497
69 (10)	790100	822900	895100	759800	815455
103.4 (15)	97700	88500	83200	88800	89401

Test data also reveal that with an increasing bandwidth of the loading (i.e. by going from the broad-band to the bimodal spectral density, see Table 9.3), the number of cycles to failure substantially increases as well. This fact becomes more evident if we consider also experimental data on narrow-band loadings reported in [Sarkani et al. 1994], not reported here. The obvious reason is that a broad-band loading contains many high-frequency low-amplitude cycles that contribute very little to the fatigue damage, but do increase the cycle count significantly.

Note however that by dividing the number of cycles to failure by the peak rate v_p (i.e. the intensity of counted cycles) drastically reduces the time to failure as the bandwidth increases. It means that the high frequency components are important because they accelerate the rate of fatigue damage accumulation. This also supports the hypothesis that under variable amplitude loadings there is no fatigue limit.

Further examination of the results also indicates that, as expected, the loading non-normality significantly modifies the fatigue life. A reduction in fatigue life for loading with kurtosis value higher than 3 (e.g. $\gamma_4 = 5$) and the same RMS level is observed for all classes of loadings investigated. The reverse is true for non-Gaussian loadings with a kurtosis less than 3 (e.g. $\gamma_4 = 2$).

9.3. ANALYTICAL FORMULAS FOR LIFETIME PREDICTION

In this section we provide the analytical formulas for estimating the number of cycles to failure. If the S-N curve is $s^k N = C$, the expected fatigue damage in time T, expressed as a function of the number of counted cycles, is (see Chapter 3):

$$\overline{D}(\overline{N}) = \overline{N} \frac{E[s^k]}{C} \tag{9.7}$$

where $\overline{N} = E[N(T)]$ is the average number of cycles counted in time period T and:

$$\Delta\overline{D} = \frac{E[s^k]}{C} \tag{9.8}$$

is the expected damage per cycle. Note that for the rainflow count, in stationary loadings, it is $\overline{N} = v_p T$, being v_p the intensity of peaks.

Fracture occurs when the total damage reaches a critical damage value D_{cr}, often set equal to unity, consequently setting $\overline{D}(\overline{N}_f) = 1$ gives the expected number of cycles to failure as:

$$\overline{N}_f = \frac{1}{\Delta\overline{D}} = \frac{C}{E[s^k]} \tag{9.9}$$

9.3.1. Deterministic rainflow analysis

The deterministic rainflow analysis (i.e. rainflow count and linear rule applied on a simulated loading) gives an estimate of the number of cycles to failure.

Let $z(t)$ be a simulated loading of duration T and denote as $D(N)$ the total rainflow damage resulting from the rainflow analysis. By dividing total damage for $N(T)$, the number of cycles counted by the rainflow method, we get an estimate of the average rainflow damage per cycle:

$$\Delta\hat{D} = \frac{D(N)}{N} \tag{9.10}$$

and similarly to Eq. (9.9), an estimate of the number of cycles to failure is:

$$N_{sim} = \frac{1}{\Delta\hat{D}} \tag{9.11}$$

The quantity N_{sim} is clearly a random variable. The subscript indicates that the estimate derives from simulations; we shall also use a different superscript to distinguish the results of the rainflow analysis from new simulations performed in this work (we will use N_{sim}^{new}), from those presented in [Kihl et al. 1995, Sarkani et al. 1996] (we shall use instead N_{sim}^{old}).

9.3.2. Narrow-band approximation

The expected damage in time T is given as (see Chapter 4):

$$\overline{D}_{NB}(T) = \frac{v_0 T}{C} \left(\sqrt{2}\,\sigma\right)^k \Gamma\left(1 + \frac{k}{2}\right) \tag{9.12}$$

where v_0, the intensity of mean upcrossings, is taken as the intensity of counted cycles, and σ is the variance of the load. Since $\overline{N} = v_0 T$ is the average number of cycles counted in time T, the expected damage per cycle is then:

$$\Delta \overline{D}_{NB} = \frac{\left(\sqrt{2}\sigma\right)^k}{C} \Gamma\left(1+\frac{k}{2}\right) \tag{9.13}$$

and finally the expected number of cycles to failure is:

$$\overline{N}_{NB} = \frac{1}{\overline{D}_{NB}} = \frac{C}{\left(\sqrt{2}\,\sigma\right)^k \Gamma\left(1+\frac{k}{2}\right)} \tag{9.14}$$

Note that for a given S-N curve, $\overline{N}_{NB}$ only depends on the variance σ.

9.3.3. Gaussian prediction

The estimation of the rainflow fatigue damage under the Gaussian hypothesis is based on the TB method illustrated in Chapter 4.

The explicit formula for the expected rainflow damage in time T is:

$$\overline{D}_{RFC}^{G}(T) = \left[b_{app} + (1 - b_{app})\alpha_2^{k-1}\right]\overline{D}_{NB}(T) \tag{9.15}$$

where b_{app} is the approximate weighting factor (see Chapter 4). It is more convenient to rewrite the above formula by underlying the expected number of rainflow counted cycles, $\overline{N} = v_p T$:

$$\overline{D}_{RFC}^{G} = \rho_{TB}\, \alpha_2\, \frac{v_p T}{C} \left(\sqrt{2}\sigma\right)^k \Gamma\left(1+\frac{k}{2}\right) \tag{9.16}$$

where we have set for simplicity $\rho_{TB} = b_{app} + (1 - b_{app})\alpha_2^{k-1}$ and we have also used the fact that $v_0 = \alpha_2 v_p$. The expected damage per cycles then writes:

$$\Delta \overline{D}_{RFC}^{G} = \frac{\rho_{TB}\, \alpha_2}{C} \left(\sqrt{2}\sigma\right)^k \Gamma\left(1+\frac{k}{2}\right) \tag{9.17}$$

and accordingly the expected number of cycles to failure is:

$$\overline{N}_G = \frac{1}{\Delta \overline{D}_{RFC}^{G}} = \frac{C}{\rho_{TB}\, \alpha_2 \left(\sqrt{2}\sigma\right)^k \Gamma\left(1+\frac{k}{2}\right)} = \frac{\overline{N}_{NB}}{\rho_{TB}\, \alpha_2} \tag{9.18}$$

9.3.4. Markov approach

The distribution of rainflow cycles can be estimated under the Markov chain approximation for the sequence of maxima and minima [Frendhal and Rychlik 1993, Olagnon 1994]. The class of random processes for which this assumption is valid is very wide and covers many types of real loading processes.

The quantities relevant for determining the expected rainflow matrix $\mathbf{Q}^{\text{rfc}}$ are the min-max, $\mathbf{P}$, and max-min, $\hat{\mathbf{P}}$, Markov transition matrices. If the random loading is assumed to be time-reversible, then it is $\hat{\mathbf{P}} = \mathbf{P}^T$.

The Markov method developed by Frendhal and Rychlik [Frendhal and Rychlik 1993] is implemented in the WAFO routine `mctp2rfm`.

If the Markov transition matrices are not known in advance, they can be estimated from a simulated or measured load. We firstly compute the observed min-max, $\mathbf{F}$, and max-min, $\hat{\mathbf{F}}$, matrices, which store the number of transitions from a maximum to the following minimum, and from a minimum to the following maximum (see Chapter 2). If the loading is time-reversible, only the min-max $\mathbf{F}$ matrix is needed.

To obtain an estimate of the expected min-max matrix $\mathbf{Q}$, the observed min-max matrix $\mathbf{F}$ is smoothed using a 2-dimensional kernel smoothing[*] (if the max-min matrix $\hat{\mathbf{F}}$ is also used, we can improve the estimate by smoothing the sum $\mathbf{F}' = \mathbf{F} + \hat{\mathbf{F}}^T$). The expected max-min matrix is $\hat{\mathbf{Q}} = \mathbf{Q}^T$

The min-max and max-min transition matrices $\mathbf{P}$ and $\hat{\mathbf{P}}$ are obtained from the corresponding expected min-max and max-min transition matrices $\mathbf{Q}$ and $\hat{\mathbf{Q}} = \mathbf{Q}^T$, respectively, by normalising the sum of each row to 1.

For modelling loads that are not time-reversible the min-max transition matrix $\mathbf{P}$ is obtained from the min-max matrix $\mathbf{F}$, and the max-min transition matrix $\hat{\mathbf{F}}$ is obtained from the max-min matrix $\hat{\mathbf{F}}$. This can be done in the same way as described above.

Once the expected rainflow matrix $\mathbf{Q}^{\text{rfc}}$ is determined, the expected rainflow fatigue damage per cycle is computed and then the expected number of cycles to failure is calculated as in Eq. (9.11), that will be denoted as N_{mkv}.

9.4. NUMERICAL SIMULATIONS AND RESULTS

In order to check the accuracy of the theoretical methods for lifetime predictions, we perform new numerical simulations, according to the details of experimental tests as illustrated in section 9.2.

These simulations aim to reproduce as close as possible the random loadings used in experiments, in order to both verify the results of rainflow analysis presented in [Kihl et al. 1995, Sarkani et al. 1996] and also to check the accuracy of analytical methods for lifetime predictions.

---

[*] The WAFO routine `smoothcmat` is used.

Results from these new simulations will also be compared with data from experimental tests. In the following, we will summarise the entire procedure that has been adopted in the present study.

9.4.1. Numerical simulations

First, we select a spectral density $W_x(\omega)$ (broad-band or bimodal), a value for the RMS (e.g. 51.7, 69 or 103.4 MPa) and a value for the kurtosis (e.g. 2 or 5). Then, we simulate a Gaussian time history $x(t)$ and we scale to RMS, so to have a standard loading $\tilde{x}(t)$ with unit RMS (i.e. $\sigma_{\tilde{x}} = 1$). Then, we transform $\tilde{x}(t)$ to non-Gaussian standard loading $\tilde{z}(t)$ (having $\sigma_{\tilde{z}} = 1$), by using parameters listed in Table 9.2 (loading $\tilde{z}(t)$ will have the chosen value of kurtosis). Finally, we rescale loading $\tilde{z}(t)$ to the correct RMS level, so to get the non-Gaussian loading $z(t)$ having the selected values of RMS and kurtosis.

The complete procedure can be summarised in the following scheme:

S1) Set parameters of simulation:
   a) select a spectral density $W_x(\omega)$ (broad-band or bimodal);
   b) select the RMS level (51.7, 69 or 103.4 MPa);
   c) select kurtosis value (2 or 5);
   d) take parameters (β, n, C) to define transformation $G(\cdot)$.

S2) Simulate Gaussian and non-Gaussian loading:
   a) simulate a Gaussian loading $x(t)$ (σ_x = RMS);
   b) scale $x(t)$ to get Gaussian the standard loading $\tilde{x}(t)$ (with $\sigma_{\tilde{x}} = 1$);
   c) transform to non-Gaussian (standard) loading $\tilde{z}(t)$ ($\sigma_{\tilde{z}} = 1$) by Eq. (9.4);
   d) scale $\tilde{z}(t)$ to correct variance σ_z to get non-Gaussian loading $z(t)$.

Step **S2**.a uses a standard frequency-domain simulation technique, while in Step **S2**.b each value of $x(t)$ is divided to the chosen RMS value. In Step **S2**.d, we finally get the non-Gaussian loading $\tilde{z}(t)$, having the desired values of RMS and kurtosis. In conclusion, for each spectral density $W_x(\omega)$ (broad-band or bimodal), we generate two random loadings, a Gaussian $x(t)$ and non-Gaussian $z(t)$ one.

In Table 9.6 and Table 9.7 we compare the time-domain characteristics of simulated loadings with their corresponding values in frequency-domain [Benasciutti and Tovo 2006b]. The base Gaussian loading $x(t)$ and the transformed non-Gaussian loading $z(t)$ have virtually the same estimated RMS value $\hat{\sigma}_x$ and $\hat{\sigma}_z$ (also equal to the theoretical value given in the first column) and the irregularity factor IF, as expected. The α_2 bandwidth parameter is computed from the spectral density: for the Gaussian loading, it is estimated from $W_x(\omega)$ and it is listed in Table 9.1, for the non-Gaussian loading, the spectral density $W_z(\omega)$ is estimated from $z(t)$ data. For the Gaussian loading it is $IF = \alpha_2$, as expected (the same is not true for the non-normal loading).

Table 9.6: Comparison of spectral parameters with time-domain parameters (kurtosis = 2).

RMS (MPa)	Gaussian data, $x(t)$				non-Gaussian data, $z(t)$			
	$\hat{\sigma}_x$	IF	α_2	γ_4	$\hat{\sigma}_z$	IF	α_2	γ_4
broad-band								
51.7	51.7	0.850	0.84	3.0	51.6	0.85	0.505	2.00
69	69	0.850	0.84	2.9	69	0.85	0.504	1.99
103.4	103.	0.850	0.84	3.0	103.	0.85	0.505	2.00
bimodal								
51.7	51.8	0.394	0.38	3.0	51.8	0.39	0.395	2.01
69	69	0.394	0.38	2.9	69	0.39	0.395	2.00
103.4	103.	0.394	0.38	3.0	103.	0.39	0.396	2.00

Table 9.7: Comparison of spectral parameters with time-domain parameters (kurtosis = 5).

RMS (MPa)	Gaussian data, $x(t)$				non-Gaussian data, $z(t)$			
	$\hat{\sigma}_x$	IF	α_2	γ_4	$\hat{\sigma}_z$	IF	α_2	γ_4
broad-band								
51.7	51.7	0.850	0.84	3.0	51.6	0.850	0.74	5.02
69	69	0.850	0.84	2.9	69	0.850	0.74	4.94
103.4	103.	0.850	0.84	3.0	102.	0.850	0.74	4.97
bimodal								
51.7	51.8	0.394	0.38	3.0	51.9	0.394	0.39	5.06
69	69	0.394	0.38	2.9	69	0.394	0.39	4.97
103.4	103.	0.394	0.38	3.0	103.	0.394	0.39	5.02

9.4.2. Procedure of analysis

In the deterministic analysis, cycles are extracted from the simulated loading by applying the rainflow counting method. The observed fatigue loading spectrum is computed from the set of rainflow counted cycles, while the fatigue damage is calculated using the linear damage rule. The fatigue lifetime follows from the rainflow damage as indicated in Eq. (9.11).

The main steps of analysis are as follows:

D1) Deterministic analysis (rainflow analysis on simulated loading $z(t)$ or $x(t)$):

   1r) apply the rainflow count to $z(t)$ (or $x(t)$);

   2r) compute rainflow cumulative $\hat{F}_{Z,\mathrm{RFC}}(s)$ (or $\hat{F}_{X,\mathrm{RFC}}(s)$);

   3r) compute number of cycles to failure $N_{\mathrm{sim}}^{\mathrm{new}}$.

D2) Markov analysis on simulated loading:

    1m) compute the min-max $\mathbf{F}$ and max-min $\hat{\mathbf{F}}$ matrices;

    2m) smooth to get the expected min-max $\mathbf{Q}$ and max-min $\hat{\mathbf{Q}}$ matrices;

    3m) normalise to get the min-max $\mathbf{P}$ and max-min $\hat{\mathbf{P}}$ transition matrices;

    4m) calculate the expected rainflow matrix $\mathbf{Q}^{\mathrm{rfc}}$ (routine `mctp2rfm`);

    5m) compute expected cycles to failure N_{mkv}.

The rainflow analysis may be applied to the Gaussian loading $x(t)$ as well. The Markov method has been included in the deterministic part of the analysis, since it requires the determination of the observed min-max and max-min matrices.

The sequence of the theoretical analysis can be summarised as:

T1) Gaussian analysis (as if $Z(t)$ were Gaussian):

    1g) estimate spectral density $W_Z(\omega)$ of $Z(t)$;

    2g) compute α_1, α_2 spectral parameters and b_{app} coefficient, Eq. (3.77);

    3g) estimate rainflow Gaussian distribution $H^{\mathrm{G}}_{Z,\mathrm{RFC}}(z_{\mathrm{p}},z_{\mathrm{v}})$ for $Z(t)$;

    4g) compute expected Gaussian rainflow cumulative $F^{\mathrm{G}}_{Z,\mathrm{RFC}}(s)$;

    5g) compute expected cycles to failure N_{G} (Gaussian).

T2) non-Gaussian analysis (treat $Z(t)$ as non-Gaussian):

    1ng) refers to the underlying Gaussian process $x(t)$;

    2ng) take spectral density $W_X(\omega)$ and estimate rainflow Gaussian distribution $H^{\mathrm{G}}_{X,\mathrm{RFC}}(x_{\mathrm{p}},x_{\mathrm{v}})$, as in Steps 2g-3g;

    3ng) transform to non-Gaussian distribution $H^{\mathrm{nG}}_{Z,\mathrm{RFC}}(z_{\mathrm{p}},z_{\mathrm{v}})$ for $Z(t)$;

    4ng) compute expected non-Gaussian rainflow cumulative $F^{\mathrm{nG}}_{Z,\mathrm{RFC}}(s)$;

    5ng) compute expected cycles to failure N_{nG} (non-Gaussian).

In the **T1** analysis $z(t)$ is treated as if it were Gaussian, i.e. we neglect its deviation from the normal behaviour. In order to estimate the rainflow cycle distribution we need to know its spectral density, which is estimated from $z(t)$ data[*].

In the case that Step **T1** is applied to a Gaussian loading $x(t)$, its spectral density is represented by the broad-band or bimodal spectrum and does not need to be estimated.

Note that in the non-Gaussian analysis applied to $z(t)$, the underlying Gaussian loading process is clearly $x(t)$, being these two loadings related by the transformation $G(\cdot)$.

---

[*] The WAFO routine `dat2spec` is used.

9.5. DISCUSSION OF RESULTS

We first compare the distribution of rainflow cycles, in terms of cumulative loading spectra, normalised to unit time (i.e. cumulated cycles/sec). The rainflow cumulative, $\hat{F}_{RFC}$, as observed from a simulated loading ($x(t)$ or $z(t)$) is compared with the expected rainflow cumulative, as estimated from the spectral density of the process.

First we refer to the Gaussian case, i.e. kurtosis equal to 3. The observed rainflow cumulative is derived from simulated Gaussian $x(t)$ data as described in Step **D1**. Then, the Steps listed in phase **T1** are applied considering spectral density $W_X(\omega)$ (broad-band or bimodal) and spectral parameters listed in Table 9.1. The expected Gaussian rainflow cumulative $F_{X,RFC}^G$ instead is computed from the loading spectral density, as described in **T1** steps. In Figure 9.4 we compare the estimated with the observed rainflow cumulative, for both broad-band and bimodal spectra, with a RMS value equal to 69 MPa (10 ksi), since plots for other RMS values are quite similar. There is in general good agreement between the two loading spectra, indicating the correctness of the method used. The expected loading spectrum allows us to extrapolate to high amplitudes, corresponding to low-probability events (i.e. cycles) not occurring in the simulated loading. Subsequent figures refer instead to the non-Gaussian case, comparing the rainflow cumulative $\hat{F}_{Z,RFC}$ observed in non-normal loading $z(t)$ with the theoretical loading spectrum $F_{Z,RFC}^{nG}$, estimated by including the non-normality of the loading into the rainflow cycle distribution. The expected rainflow cumulative $F_{Z,RFC}^G$ estimated without including the non-normality of the loading is also shown.

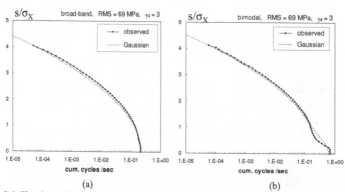

Figure 9.4: The observed rainflow cumulative is compared with the estimated Gaussian rainflow cumulative. RMS 69 MPa (10 ksi). (a) Broad-band and (b) Bimodal spectral density.

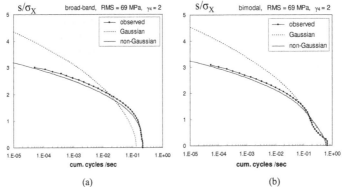

(a)   (b)

Figure 9.5: The observed rainflow cumulative compared with the estimated Gaussian and non-Gaussian cumulative. RMS 69 MPa (10 ksi), kurtosis = 2. (a) broad-band, (b) bimodal spectrum.

Figure 9.5 refers to a platykurtic loading, having a kurtosis equal to 2. On the contrary, Figure 9.6 shows the results for a leptokurtic loading, having a kurtosis equal to 5. In this case, the non-normal process has higher probability towards largest cycles, leading to greater damaging character of the observed rainflow cumulative.

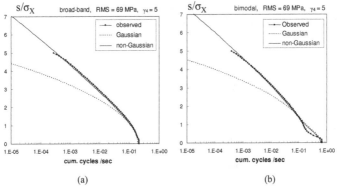

(a)   (b)

Figure 9.6: The observed rainflow cumulative compared with the estimated Gaussian and non-Gaussian cumulative. RMS 69 MPa (10 ksi), kurtosis = 5. (a) broad-band, (b) bimodal spectrum.

Only the non-Gaussian estimation seems to agree with the observed rainflow cumulative, whereas the Gaussian estimation, neglecting the non-normal character of the loading, underestimates the probability of the largest cycles.

Once we have the statistical distribution of rainflow cycles, we can calculate the fatigue damage under the linear damage accumulation rule.

The tables below report the comparison of fatigue lives amongst numerical simulations, theoretical predictions and experimental tests, while in Figure 9.7 through Figure

9.13 we sketch the same results. The abscissa is the estimated lifetime and the ordinate is the lifetime from experiments, reported both as the geometric mean that as result of single tests (which are available only for kurtosis = 5). As usual, the straight line at 45 degrees represents the perfect matching between theoretical prediction and experimental result. Thus, points lying above indicate a more conservative prediction (i.e. the estimated life is lower than observed), whereas points below indicates a less conservative prediction (i.e. the estimated life is higher than observed).

The geometric mean of the lifetimes resulting in experimental, N_{exp}, listed in the second column of each table, are extracted from the last column of Table 9.3, Table 9.4 and Table 9.5, and they will be used in the following for calculating the deviations occurring between simulations and theoretical predictions.

The result of rainflow analysis is reported in the third and fourth column of each table. The lifetimes indicated as N_{sim}^{old} are taken from [Kihl et al. 1995, Sarkani et al. 1996] and refer to simulated loadings used for the experiments, whereas the lifetimes labelled as N_{sim}^{new} result from new simulations.

The lifetime as predicted by the narrow-band approximation is given under the label N_{nb}; it disregards the loading bandwidth, being only dependent on the RMS value.

Finally, in the last three columns we report the predictions from the Gaussian, non-Gaussian assumption and the Markov method, respectively.

As expected, the difference between the results of rainflow analysis, N_{sim}^{old} and N_{sim}^{new}, is always less than 2.5% for all cases. Thus in the following, all comparisons involving simulations will refer to N_{sim}^{new} data.

Table 9.8: Fatigue lifetime for Gaussian loadings (kurtosis = 3)

RMS MPa (ksi)	EX. N_{exp}	RFA (old data) N_{sim}^{old}	RFA (new N_{sim}^{new}	NB N_{nb}	G N_{G}	nG N_{nG}	MKV N_{mkv}
broad-band							
51.7 (7.5)	1410491	1615900	1611045	1288382	1644031	–	1553645
69 (10)	515124	641800	637296	511672	652915	–	617019
103.4 (15)	188718	174600	174867	139232	177666	–	167898
bimodal							
51.7 (7.5)	3932840	4460100	4355360	1288382	4685387	–	5408769
69 (10)	2027726	1771400	1728315	511672	1860767	–	2148053
103.4 (15)	386746	482000	473485	139232	506336	–	584510

EX. = Experimental; RFA = rainflow analysis; NB = narrow-band approximation;
 G = Gaussian prediction; nG = non-Gaussian prediction; MKV = Markov method (Frendhal &

Table 9.9: Fatigue lifetime for non-Gaussian loadings (kurtosis = 2)

RMS MPa (ksi)	EX. N_{exp}	RFA (old data) $N_{\text{sim}}^{\text{old}}$	RFA (new $N_{\text{sim}}^{\text{new}}$	NB N_{nb}	G N_{G}	nG N_{nG}	MKV N_{mkv}
broad-band							
51.7 (7.5)	2431900	–	2300332	1288382	2704898	2332581	1973733
69 (10)	888000	–	904634	511672	1075850	926359	777308
103.4 (15)	325350	–	250315	139232	292280	252075	214807
bimodal							
51.7 (7.5)	6780100	–	6337500	1288382	4649563	6566613	5994109
69 (10)	3496000	–	2504725	511672	1847774	2607923	2368059
103.4 (15)	666700	–	693010	139232	501218	709635	655382

EX. = Experimental; RFA = rainflow analysis; NB = narrow-band approximation;
G = Gaussian prediction; nG = non-Gaussian prediction; MKV = Markov method (Frendhal & Rvchlik)

Table 9.10: Fatigue lifetime for non-Gaussian loadings (kurtosis = 5)

RMS MPa (ksi)	EX. N_{exp}	RFA (old data) $N_{\text{sim}}^{\text{old}}$	RFA (new $N_{\text{sim}}^{\text{new}}$	NB N_{nb}	G N_{G}	nG N_{nG}	MKV N_{mkv}
broad-band							
51.7 (7.5)	853573	1085800	1088288	1288382	1839681	1106621	1009381
69 (10)	316407	431200	430653	511672	730464	437244	398423
103.4 (15)	44168	117300	119432	139232	198786	120142	110600
bimodal							
51.7 (7.5)	2610497	2838200	2831655	1288382	4663236	3111311	2917747
69 (10)	815455	1127100	1127427	511672	1850284	1232422	1154524
103.4 (15)	89401	306800	312099	139232	502490	339105	320710

EX. = Experimental; RFA = rainflow analysis; NB = narrow-band approximation;
G = Gaussian prediction; nG = non-Gaussian prediction; MKV = Markov method (Frendhal & Rvchlik)

A good agreement is generally observed between the results of rainflow analysis ($N_{\text{sim}}^{\text{old}}$ and $N_{\text{sim}}^{\text{new}}$) and experimental data N_{exp} for both Gaussian and non-Gaussian load-ings. Under Gaussian loadings the maximum deviation is less than 25%, occurring for a RMS value equal to 69 MPa (10 ksi). Under non-Gaussian loadings, the deviation is less than 38%, except the non-normal loadings with high RMS levels (103.4 MPa) and kurtosis 5. The large disagreement may be attributed to the fact that only these loadings contain many cycles having large amplitudes near the yield strength of the base steel plate. It is possible that these large cycles' amplitudes accelerate the rate of fatigue dam-age accumulation.

Amongst all theoretical methods for lifetime assessment, the lifetime prediction N_{nb} from the narrow-band approximation is highly conservative, always with shorter fatigue lives than those observed in both simulations and experiments. In the Gaussian case (see Table 9.8) the deviation is small (about 20%) for the broad-band spectrum, and even as large as 70% for the bimodal spectral density (confirming that the bimodal spectrum is actually more wide-banded than the broad-band spectrum).

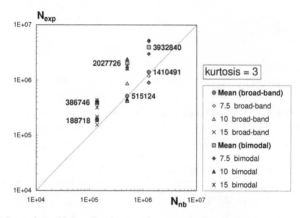

Figure 9.7: Comparison of fatigue lives between the narrow-band approximation and experimental test results (given also as figures). Gaussian case (kurtosis = 3).

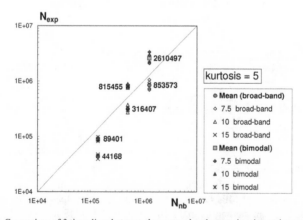

Figure 9.8: Comparison of fatigue lives between the narrow-band approximation and experimental test results (given also as figures). non-Gaussian case (kurtosis = 5).

176

Compared to the Gaussian case, the non-Gaussian situation may give different deviations, depending on the kurtosis value and to the type of spectrum. As expected, in non-normal platykurtic loadings (kurtosis equal to 2), larger deviations occur, about 44% for the broad-band spectrum and about 80% for the bimodal spectrum (see Table 9.9). On the contrary, in non-normal leptokurtic loadings (kurtosis equal to 5), smaller deviations are present, about 18% for the broad-band spectrum and about 55% for the bimodal spectrum. Nevertheless, the prediction made by the narrow-band approximation is only valid under the Gaussian hypothesis.

The lifetime predictions according to the TB method are listed in the sixth and seventh column in Table 9.8 through Table 9.10: the label N_G indicates the prediction under the Gaussian hypothesis for the loading (i.e. as if $z(t)$ were Gaussian), whereas N_{nG} includes the non-normality.

As can be seen in Table 9.8 (the N_{nG} data are clearly not reported), the method seems to work very well. The maximum deviation from data resulting in simulations (i.e. N_{sim}^{new}) is about 8% with the bimodal spectral density (i.e. the most wide-banded spectrum), and as large as 2% for the broad-band spectrum. At the same time, prediction diverges from experiments of about 30% for both spectral densities, similarly to the difference observed between numerical simulations and experiments.

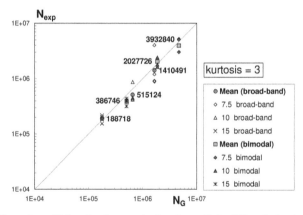

Figure 9.9: Comparison of fatigue lives between the Gaussian prediction (TB method) and experimental test results (given also as figures). Gaussian case (kurtosis = 3).

In the non-Gaussian case the prediction by Gaussian method is clearly in defect, since it does not take into account the non-normality of the loading. Referring to results observed in simulations, in a platykurtic loading (kurtosis equal to 2) the predicted fatigue life is greater for the broad-band spectrum (of about positive 18%) and smaller (of about negative 26%) for the bimodal spectrum. More pronounced differences are observed respect to experimental results. In a leptokurtic loading (kurtosis equal to 5), the

177

Gaussian prediction is always not conservative, leading to fatigue lives longer than observed (deviation of about positive 65% for both spectral densities).

These large discrepancies are completely cancelled if the rainflow cycle distribution is estimated by the non-Gaussian method, N_{nG}. In both non-Gaussian cases that have been examined, the difference between the theoretical prediction N_{nG} and the result from simulations, N_{sim}^{new}, is very small, the maximum deviation being generally about positive 5% in all cases, except the bimodal spectrum with kurtosis equal to 5, where it arrives at 10%.

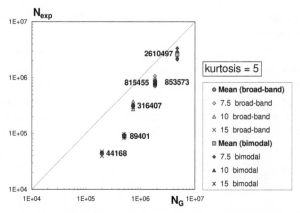

Figure 9.10: Comparison of fatigue lives between the Gaussian prediction (TB method) and experimental test results (given also as figures). non-Gaussian case (kurtosis = 5).

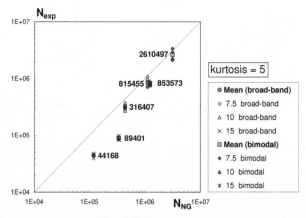

Figure 9.11: Comparison of fatigue lives between the non-Gaussian prediction (TB method) and experimental test results (given also as figures). non-Gaussian case (kurtosis = 5).

Finally, we compare the lifetimes as predicted by the Markov method, N_{mkv}

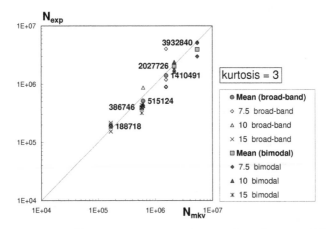

Figure 9.12: Comparison of fatigue lives between the Markov method (Frendhal and Rychlik) and experimental test results (given also as figures). Gaussian case (kurtosis = 3).

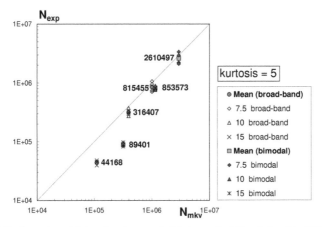

Figure 9.13: Comparison of fatigue lives between the Markov method (Frendhal and Rychlik) and experimental test results (given also as figures). Gaussian case (kurtosis = 5).

In the Gaussian case, the difference is about negative 4% for the broad-band spectrum and about positive 24% for the bimodal spectrum. In the non-Gaussian case, the difference decreases: for a kurtosis equal to 2, the deviation is about negative 14% (broad-band spectrum) and negative 5% (bimodal spectrum), and for a kurtosis equal to 5 it becomes as large as negative 7% (broad-band case) and positive 3% (bimodal case).

Based on previous results, some important conclusions may be drawn and summarised according to the following points:

- the quite good agreement between the rainflow analysis results N_{sim}^{new} and N_{sim}^{old} should be expected, since both results follow from simulations;
- agreement between a theoretical model and the rainflow analysis indicates good accuracy of the model in predicting the rainflow cycle distribution;
- the disagreement between both the rainflow analysis and the theoretical model in respect of experimental data may indicate the inadequacy of the linear damage accumulation assumption.

We can conclude that the TB method gives good predictions in both the Gaussian and the non-Gaussian case, in the last case including proper loading non-normality into the rainflow cycle distribution. The method is quite simple, requiring only the definition of the transformation $G(\cdot)$ and the knowledge of the Gaussian spectral density $W_X(\omega)$.

Quite good results are also given by the Markov method, which may be indifferently applied both Gaussian and non-Gaussian cases, only requiring the knowledge of the one-step min-max transition matrix.

In any case, results confirm that special attention should always be paid when evaluating the fatigue lifetime of a structure under non-Gaussian loading.

Chapter 10

CONCLUSIONS

This book addresses the fatigue analysis of irregular loadings, as those observed in structure and mechanical components in service. Attention mainly focuses on the distribution of rainflow cycles and the consequent fatigue damage calculated under the Palmgren-Miner linear damage rule.

The irregular loading is modelled as a stationary random process, described in frequency-domain by a spectral density function. A set of spectral moments and bandwidth parameters are used to characterise the frequency distribution of a given spectral density. General relations amongst spectral parameters are given (see Appendix A).

Two types of stationary processes are investigated, i.e. Gaussian and non-Gaussian processes. Some basic concepts about the fatigue analysis of random processes are discussed in Chapter 3. Two important descriptor of the distribution of counted cycles are introduced, i.e. the count intensity $\mu(u, v)$ and the joint probability density function $h(u, v)$. General properties of counting methods are introduced (i.e. crossing-consistency and completeness), as well as some general properties of the $h(u, v)$ density. Since in Chapter 2 the rainflow count is recognised as the best counting procedure, attention mainly focuses on the distribution of rainflow cycles, $h_{\mathrm{RFC}}(u, v)$, and on the rainflow fatigue damage intensity $\overline{D}_{\mathrm{RFC}}$ calculated under the linear damage rule.

The fatigue analysis of Gaussian processes is presented in Chapter 4, where attention focuses on broad-band processes, i.e. processes with a spectral density extending over a wide range of frequencies. Several spectral methods for the rainflow fatigue analysis are reviewed: the narrow-band approximation, the Wirsching-Light correction formula, the Dirlik's approximate amplitude density, the Zhao-Baker weighted amplitude density, the Petrucci-Zuccarello approximate damage.

Further, two original methods are proposed: the TB method and the empirical-$\alpha_{0.75}$ method. In the TB method rainflow damage is estimated as a weighted linear combination from two damage values, where a weighting factor b is introduced. The two damage values correspond to the damage provided by the narrow-band approximation and the range counting method, respectively; the range counting damage is estimated by an approximate formula. The weighting factor b is assumed to depend on the process spectral density. Since no information is available about this kind of dependence, an approximate relation is searched, in which b depends on α_1 and α_2 bandwidth parameters (see Appendix B). Numerical simulations are used to calibrate this approximate relation. The empirical-$\alpha_{0.75}$ method, instead, is an approximate method which (similarly to the Wirsching-Light method) introduces a damage correction factor of the narrow-band approximation, which depends on $\alpha_{0.75}$ bandwidth parameter.

All possible analogies and differences amongst all methods cited above are evidenced and discussed. In particular, we note that some methods simply rely on fatigue damage estimation, while other also address the problem of determining the rainflow cycle distribution. For example, the narrow-band approximation and the Wirsching-Light approach, as well as the Petrucci-Zuccarello method, only estimate the rainflow damage $\overline{D}_{RFC}$, while the Dirlik and Zhao-Baker methods estimate also the marginal amplitude distribution $p_a(s)$ of rainflow cycles. The TB method, instead, is the only one method that is also able to estimate the joint distribution $h_{RFC}(u, v)$ (or the related amplitude-mean distribution $p_{a,m}(s, m)$). These differences are recognised as very important, if we want to include the mean value influence on damage computation or non-Gaussian effects.

Furthermore, the dependence on particular set of spectral moments are underlined. For example, the narrow-band approximation and the Wirsching-Light method depend on just three spectral moments (i.e. λ_0, λ_2, λ_4), while the Dirlik and the TB method introduce a further dependence on λ_1 spectral moment. The Zhao-Baker method (in its improved version) makes the rainflow damage dependent also on $\alpha_{0.75}$ bandwidth parameter.

In order to check and compare the accuracy of all methods in the fatigue damage estimation, we numerically simulate Gaussian random processes, with different combination of α_1 and α_2 bandwidth parameters. Spectral densities having simple geometries (e.g. constant, linear, parabolic) are used to this purpose.

Results of time-domain calculations (i.e. rainflow count and linear damage rule are compared with different predictions provided by all the spectral methods mentioned above). Comparisons consider both the distribution of rainflow cycles (evaluated by means of the fatigue loading spectrum) and the rainflow fatigue damage.

First of all, results from numerical simulations are used to calibrate the b weighting factor introduced in the TB method. Subsequently, the comparison of the distribution of rainflow cycles (in terms of loading spectra) is investigated for the Dirlik, Zhao-Baker and TB method (since the other methods do not give such kind of information). Finally, the comparison of the rainflow fatigue damage value is considered.

The results of numerical simulations concerning fatigue damage confirm that the rainflow damage seems to depend on both α_1 and α_2 bandwidth parameters. Therefore, improved damage estimations should also account for a correlation on (at least) four spectral moments (i.e. λ_0, λ_1, λ_2, λ_4). In addition, since more complex connexions amongst spectral parameters exist (see Appendix A), it cannot be completely neglected a possible dependence on other parameters, as $\lambda_{0.75}$ and $\lambda_{1.5}$.

Simulations also confirm that the narrow-band approximation and the Wirsching-Light correction formula both give in broad-band processes too conservative estimates of the rainflow fatigue damage. In fact, the damage provided by the narrow-band approximation is only an upper bound for the rainflow damage.

On the contrary, the most accurate spectral methods seem to be the Dirlik, Zhao-Baker and TB methods, as well as the empirical-$\alpha_{0.75}$ method. However, the Zhao-Baker technique show some limits of applicability, i.e. it can not be applied to all our processes. Further, the empirical-$\alpha_{0.75}$ method, even if quite simple, is completely lacking of any theoretical justification, thus we suggest using it only as a first approximation of rainflow damage.

Similarly, the Dirlik method, even providing the most precise damage predictions (as already shown by data from literature), has no theoretical framework (i.e. it represents a completely approximate formula), and in addition it cannot include both the mean value influence and the effect of non-normalities. At the opposite, the TB method is as accurate as the Dirlik method, it is supported by a theoretical background, and it can provide also the distribution of rainflow cycles $h_{RFC}(u, v)$ in terms of peak u and valley v levels, thus making possible to account for mean value and non-normality influence.

Subsequently, the book investigates Gaussian random processes with a bimodal spectral density, i.e. a spectrum with two well-separated frequency components. These types of processes are important because they are often used to model the loading response measured in off-shore structures and automotive chassis components. Moreover, in bimodal random processes we know in advance that the rainflow count extracts two types of cycles: large cycles, generated by the interaction between the low- and high-frequency component, and small cycles, associated to the high-frequency component. This argument makes the rainflow fatigue analysis easier: the rainflow fatigue damage is then assumed as simply the sum of two damage contributions.

Most methods available in the literature and specifically developed for bimodal processes are based on this argument, e.g. the Jiao-Moan, the Sakai-Okamura and the Fu-Cebon method. Another method valid for bimodal processes, i.e. the single moment technique, is instead based on heuristic arguments.

In Chapter 5 we review in detail all these methods and we also propose a modification of the Fu-Cebon method, which is expected to provide improved damage predictions.

Numerical simulations of bimodal random processes are performed. Two types of bimodal spectral densities are considered, namely ideal bimodal spectra (formed by the superposition of two rectangular blocks) and more realistic two-peaked spectra. As usual, comparison is made between theoretical and time-domain rainflow damage values.

Not all methods give good predictions; for example, the Sakai-Okamura method provides poor estimations (in particular for high values of the S-N slope), which are comparable with those given by the narrow-band approximation. Similarly, the Fu-Cebon method is more accurate than the Sakai-Okamura method, even if it requires numerical integration, which cannot be achieved in some cases (i.e. the damage cannot be calculated). The modification introduced by the modified Fu-Cebon method seems to considerably improve the damage estimation.

On the contrary, quite accurate estimations are provided by the single-moment, the Jiao-Man and the TB method. In particular, the Jiao-Moan method, even if very accurate, has an applicability that does not cover all spectral densities analysed.

Finally, the fatigue analysis of Gaussian random process concludes with a simple application: a two-degrees-of freedom model moving on a irregular profile is considered, as a first approximation of a car-quarter moving on a irregular road. This simple case study shows all potentialities of frequency-domain methods: by mean of the classical frequency response analysis, the transfer functions relating the external input (i.e. the road input) to the two forces acting on the model are first determined. Then, linear analysis is used to find the spectral densities of the two forces from the spectral density of the road, which is known from literature.

The rainflow damage associated to the two forces is investigated: as usual, some spectral methods are used to compare frequency-domain results, as derived from the force spectral densities, with time-domain results given by simulations.

The results show that spectral methods can provide fast damage estimation even at the design stage: the effect on fatigue damage produced by a change in one model parameter is investigated. Spectral methods, which are in general not so accurate (e.g. the narrow-band approximation and the Wirsching-Light method), can greatly overestimate the effect on fatigue damage produced by a change of one model parameter.

In the second part of the book, attention focuses on non-Gaussian processes. In fact, even if the Gaussian hypothesis is commonly adopted by most of the methods available in literature, it is seldom verified by loading responses measured in real structures, due to non-linearities often present into real systems.

As a consequence, spectral methods valid for normal processes, when applied to such non-Gaussian responses, might produce non-conservative fatigue damage predictions. Therefore, such methods should be further developed to include non-Gaussian effects.

The works available in literature generally provide analytical methods restricted to non-Gaussian processes with a narrow-band spectral density (i.e. they are limited to narrow-band non-Gaussian loadings), even if real loads generally have broad-band spectra, and they also include the non-normal effects by the kurtosis coefficient alone.

In Chapter 7, the main theoretical aspects of the fatigue analysis of non-Gaussian random processes are presented. A time-independent transformation $G(\cdot)$ relating a non-Gaussian process $Z(t)$ to an underlying Gaussian process $X(t)$ is introduced (which corresponds to the so-called transformed Gaussian model [Rychlik et al. 1997]). The transformation depends on the degree of deviation from the Gaussian behaviour, quantified by either the skewness or the kurtosis (or both) of the non-Gaussian process. Several parametric definitions of the transformation are reviewed, as the power-law model [Sarkani et al. 1994], the exponential model [Ochi an Ahn 1994] and the Hermite model [Winterstein 1988], as well as the non-parametric definition proposed by Rychlik et al. [Rychlik et al. 1997].

By means of the transformation, a cycle (z_p, z_v) counted in the non-Gaussian domain is related to a corresponding transformed cycle (x_p, x_v) in the Gaussian domain, where the relationship is established by applying the transformation separately to peaks z_p and x_p, and valleys z_v and x_v. Since the transformation is strictly non-decreasing, both processes $Z(t)$ and $X(t)$ have the same peak-valley coupling by the rainflow method (i.e. cycle are counted at the same time instants, but they have different amplitudes and mean values).

Since peaks and valleys in a random process are random variables, the transformation establishes a general relationship between the cumulative distribution function of rainflow cycles in the non-Gaussian domain, $H_{Z,\mathrm{RFC}}^{\mathrm{nG}}(z_p, z_v)$, and that valid in the Gaussian domain, $H_{X,\mathrm{RFC}}^{\mathrm{G}}(x_p, x_v)$. The joint density of rainflow counted cycles $h_{Z,\mathrm{RFC}}^{\mathrm{nG}}(z_p, z_v)$ (accounting for non-normality) is finally obtained by simply differentiating the corresponding cumulative distribution $H_{Z,\mathrm{RFC}}^{\mathrm{nG}}(z_p, z_v)$.

The theoretical background highlights that a necessary condition for applying the method is to have all distributions (i.e. joint densities or cumulative distributions) expressed in terms of peak and valley levels. Therefore, amongst all methods reviewed for the case of Gaussian processes, only the TB method can be applied.

Data measured on a Mountain-bike on off-road tracks and data from an automotive application are used to test the new method.

As evidenced by the comparison between the estimated and the experimental loading spectra, we show how the estimated non-Gaussian distribution of cycles counted by the rainflow method is able to include the non-normal behaviour of the load. In particular, the non-Gaussian distribution is able to correctly extrapolate the experimental spectrum towards large amplitudes (i.e. largest cycles). Conversely, this is not true with the Gaussian distribution which disregards non-Gaussian effects.

Finally, the book concludes with a comparison between lifetime data obtained in experimental tests and theoretical estimations provided by different spectral methods. Experimental data taken from the literature and referring to a welded cruciform joint subjected to non-Gaussian random loadings, with two types of spectral density, is used.

The non-Gaussian TB method and the Markov approach are applied. Presented results evidenced that a good prediction is generally observed between theoretical estimation given by the TB method and lifetime data results from rainflow analysis of simulated time histories, which indicates a fairly good reliability of the method. However, some discrepancies are observed between lifetime data from simulations and from experiments, probably indicating that the simple linear damage rule cannot be so accurate in accounting for the process of damage accumulation occurring into the material under random loading.

REFERENCES

Abdullah S., Giacomin J.A., Yates J.R. (2004). A mission synthesis algorithm for fatigue damage analysis. *Proc. Inst. Mech. Eng. Part D-J. Automob. Eng.*, 218(3), 243-258.

Agerskov H., Nielsen J.A. (1999). Fatigue in steel highway bridges under random loading. *J. Struct. Eng.-ASCE*, 125(2), 152-162.

Agerskov H. (2000). Fatigue in steel structures under random loading. *J. Constr. Steel. Res.*, 53, 283-305.

Amzallag C., Gerey J.P., Robert J.L., Bahuaud J. (1994). Standardization of the rainflow counting method for fatigue analysis. *Int. J. Fatigue*, 16, 287-293.

Anthes R.J. (1997). Modified rainflow counting keeping the load sequence. *Int. J. Fatigue*, 19(7), 529-535.

ASTM Designation E 1049-85 (1985). Standard practices for cycle counting in fatigue analysis.

Atzori B., Tovo R. (1994). I metodi per il conteggio dei cicli di fatica: stato dell'arte, problemi e possibilità di sviluppo. (Counting methods for fatigue cycles: state of the art, problems and possible developments). *ATA Ingegneria Automobilistica*, 47(4), 175-183 (in Italian).

Bel Knani K., Benasciutti D., Signorini A., Tovo R. (2007). Fatigue damage assessment of a car body-in-white using a frequency-domain approach. *Int. J. Mater. Prod. Technol.*, 30(1/2/3), 172-198.

Benasciutti D., Tovo R. (2004a). Rainflow cycle distribution and fatigue damage in Gaussian random loadings. Report n. 129, Department of Engineering, University of Ferrara, Italy.

Benasciutti D., Tovo R. (2004b). Comparison of spectral methods for fatigue analysis in broad-band Gaussian random processes. Complex Systems and JCSS Fatigue Workshop, DTU, Lyngby (Copenhagen), Denmark.

Benasciutti D., Tovo R. (2005a). Spectral methods for lifetime prediction under wide-band stationary random processes. *Int. J. Fatigue*, 27(8), 867-877.

Benasciutti D., Tovo R. (2005b). Cycle distribution and fatigue damage assessment in broad-band non-Gaussian random processes. *Prob. Eng. Mech.*, 20(2), 115-127.

Benasciutti D., Tovo R. (2006a). Comparison of spectral methods for fatigue analysis in broad-band Gaussian random processes. *Prob. Eng. Mech.*, 21(4), 287-299.

Benasciutti D., Tovo R. (2006b). Fatigue life assessment in non-Gaussian random loadings. *Int. J. Fatigue*, 28(7), 733-746.

Benasciutti D., Tovo R. (2007a). On fatigue damage assessment in bimodal random processes. *Int. J. Fatigue*, 29(2), 232-244.

Benasciutti D., Tovo R. (2007b) Frequency-based fatigue analysis of non-stationary switching random loads, *Fatigue Fract. Eng. Mater. Struct.*, 30(11), 1016-1029.

Benasciutti D., Cristofori A. (2008). A frequency-domain formulation of MCE method for multiaxal random loadings. *Fatigue Fract. Eng. Mater. Struct.*, 31(11), 937-948.

Benasciutti D., Tovo R. (2010). On fatigue cycle distribution in non-stationary switching loadings with Markov chain structure. *Prob. Eng. Mech.*, 25(4), 406-418.

Berger C., Eulitz K.G., Heuler P., Kotte K.L., Naundorf H., Shuetz W., Sonsino C.M., Wimmer A., Zener H. (2002). Betriebsfestigkeit in Germany – an overview. *Int. J. Fatigue*, 24, 603-625.

Bishop N.W.M., Sherrat F. (1990). A theoretical solution for the estimation of 'rainflow' ranges from power spectral density data. *Fatigue Fract. Eng. Mater. Struct.*, 13(4), 311-326.

Bishop N.W.M. (1994). Spectral methods for estimating the integrity of structural components subjected to random loading. Handbook of Fatigue Crack Propagation in Metallic Structures, A. Carpinteri (Editor), Elsevier Science B.V., pp. 1685-1720.

Bogsjö K. (2002). Fatigue relevant road surface statistics. Master thesis 2002:E42, Mathematical Statistics, Centre for Mathematical Sciences, Lund Institute of Technology, Lund (Sweden).

Bouyssy V., Naboishikov S.M., Rackwitz R. (1993). Comparison of analytical counting methods for Gaussian processes. *Struct. Saf.*, 12, 35-57.

Braccesi C., Cianetti F., Lori G., Pioli D. (2005). Fatigue behaviour analysis of mechanical components subjected to random bimodal stress process: frequency domain approach. *Int. J. Fatigue*, 27, 335-345.

Colombi P., Doliński K. (2001). Fatigue lifetime of welded joints under random loading: rainflow cycle vs. cycle sequence method. *Prob. Eng. Mech.*, 16, 61-71.

Cristofori A., Benasciutti D., Tovo. R. (2011a). A stress invariant based spectral method to estimate fatigue life under multiaxial random loading. *Int. J. Fatigue*, 33(7), 887-899.

Cristofori A., Benasciutti D., Tovo. R. (2011b). Analogie fra metodi spettrali e criteri multiassiali nella valutazione del danno a fatica. *Proc. of the 40th Conference of the Italian Stress Analysis Association (AIAS)*, Palermo, 2001 (in Italian).

Deodatis G., Koutsourelakis P.S., Micaletti R.C. (2001). A methodology to simulate strongly non-Gaussian stochastic processes and fields. In: *Proc. of the Int. Conf. Structural Safety and Reliability* ICOSSAR 2001, Corotis et al. Editors.

Deodatis G., Micaletti R.C. (2001). Simulation of highly skewed non-Gaussian stochastic processes. *J. Eng. Mech.-ASCE*, 127(12), 1284-1295.

Dietrich C.R., Newsam G.N. (1997). Fast and exact simulation of stationary Gaussian process through circulant embedding of the Covariance matrix. *SIAM J. Sci. Comput.*, 18(4), 1088-1107.

Dirlik T. (1985) Application of computers in fatigue analysis. PhD Thesis, University of Warwick, UK.

Dodds C.J., Robson J.D. (1973). The description of road surface roughness. *J. Sound Vibr.*, 31(2), 175-183.

Dowling N.E. (1972). Fatigue failure predictions for complicated stress-strain histories. *Journal of Materials JMLSA*, 7(1), 71-87.

Downing S.D., Socie D.F. (1982). Simple rainflow counting algorithms. *Int. J. Fatigue*, 4(1), 31-40.

Dreßler K., Hack M., KrügerW. (1997). Stochastic reconstruction of loading histories from a rainflow matrix. *Zeitschrift für Angewandete Mathematik und Mechanik*, 77, 217-226.

ESDU, Fatigue life estimation under variable amplitude loading using cumulative damage calculations. *ESDU 95006*. Engineering Sciences Data Unit, London.

Fatemi A, Young L. (1997). Cumulative fatigue damage and life prediction theories: a survey of the state of the art for homogeneous materials. *Int. J. Fatigue*, 20(1), 9-34.

Frendhal M., Rychlik I. (1993). Rainflow analysis: Markov method. *Int. J. Fatigue*, 15, 265-272.

Fu T.T., Cebon D. (2000). Predicting fatigue lives for bi-modal stress spectral densities. *Int. J. Fatigue*, 22, 11-21.

Fuchs H.O., Nelson D.V., Burke M.A., Toomay T.L. (1977). *Fatigue Under Complex Loading: Analyses and Experiments*, Vol. AE-6, Wetzel R.M. editor, The Society of Automotive Engineers.

Gao Z., Moan T. (2008). Frequency-domain fatigue analysis of wide-band stationary Gaussian processes using a trimodal spectral formulation. *Int. J. Fatigue*, 30(10/11), 1944-1955.

Gobbi M., Mastinu G. (1998). Expected damage of road vehicles due to road excitation. *Vehicle System Dynamic Supplement*, 28, 778-788.

Gobbi M., Mastinu G. (2000). Considerazioni teoriche relative alla progettazione di casse di sospensioni di autoveicoli. (Theoretical considerations on the design of suspension systems). *Proceedings of the AIAS meeting "Experiences in application of loading spectra for fatigue design"*, Milan, Italy (in Italian).

Grigoriu M. (1993). On the spectral representation method in simulation. *Prob. Eng. Mech.*, 8, 75-90.

Gurley K.R., Kareem A., Tognarelli M.A. (1996). Simulation of a class of non-normal random processes. *Int. J. Non-Linear Mech.*, 31(5), 601-617.

Gurley K.R., Kareem A. (1999). Applications of Wavelet transforms in earthquake, wind, and ocean engineering. *Eng. Struct.*, 21(2), 149-167.

Haiba M., Barton D.C., Brooks P.C., Levesley M.C. (2002). Review of life assessment techniques applied to dynamically loaded automotive components. *Comput. Struct.*, 80, 481-494.

Halfpenny A. (1999). A frequency domain approach for fatigue life estimation from Finite Element Analysis. International Conference on Damage Assessment of Structure (DAMAS 99), Dublin.

Holmes J.D. (2002). Fatigue life under along-wind loading – closed-form solutions. *Eng. Struct.*, 24, 109-114.

Hong N. (1991). A modified rainflow counting method. *Int. J. Fatigue*, 13(6), 465-469.

Hu B., Shiehlen W. (1997). On the simulation of stochastic processes by spectral representation. *Prob. Eng. Mech.*, 12(2), 105-113.

ISO (1995). Mechanical vibration - Road surface profile – Reporting measured data. ISO 8608:1995(E).

189

Jha A.K., Winterstein S. (2000). Stochastic fatigue damage accumulation due to nonlinear ship loads. *J. Offshore Mech. Arct. Eng. Trans. ASME*, 122, 253-259.

Jiao G., Moan T. (1990). Probabilistic analysis of fatigue due to Gaussian load processes. *Prob. Eng. Mech.*, 5(2), 76-83.

Johannesson P., Lindgren G., Rychlik I. (1995). Rainflow modelling of random vehicle fatigue loads. *ITM (Swedish Inst. of Appl. Mathematics)*, Report 1995:5.

Johannesson P. (1998). Rainflow cycles for switching processes with Markov structure. *Probab. Eng. Inform. Sci.*, 12, 143-175.

Johannesson P. (1999). Rainflow analysis of switching Markov loads. PhD thesis, Mathematical Statistics, Centre for Mathematical Sciences, Lund Institute of Technology, Lund (Sweden).

Johannesson P., Thomas J. (2001). Extrapolation of rainflow matrices. *Extremes*, 4(3), 241-262.

Johannesson P., Thomas J., de Maré (2002). Extrapolation and scatter of test track measurements. *Fatigue 2002* (Editor A.F. Blom), Stockholm, Sweden (June 2002).

Johannesson P. (2004). Extrapolation of load histories and spectra. *Proceedings of ECF15 (European Conference on Fracture)*, Stockholm, Sweden.

Kececioglu D.B., Jiang M.X., Sun F.B. (1998). A unified approach to random-fatigue reliability quantification under random loading. *Proceedings of Annual Reliability and Maintainability Symposium IEEE*, 308-313.

Kihl D.P., Sarkani S., Beach J.E. (1995). Stochastic fatigue damage accumulation under broadband loadings. *Int. J. Fatigue*, 17(5), 321-329.

Kim J.J., Kim H.Y. (1994). Simple method for evaluation of fatigue damage of structures in wide-band random vibrations. *Proc. Inst. Mech. Eng. Part C-J. Mech. Eng. Sci.*, 208(C1), 65-68.

Košút J. (2002). History influence exponent in cumulative fatigue damage determined using two-step loading experiments. *Fatigue Fract. Eng. Mater. Struct.*, 25, 575-586.

Košút J. (2004). Quadratic damage rule in random loading case. *Fatigue Fract. Eng. Mater. Struct.*, 27(8), 679-700.

Kowalewski J. (1966). On the relationship between component life under irregularly fluctuating and ordered load sequences, *MIRA Translations* n. 43/66 (part 1), n. 60/66 (part 2).

Krenk S., Gluver H. (1989). A Markov matrix for fatigue load simulation and rainflow range evaluation. *Struct. Saf.*, 6(2-4), 247-258.

Kukkanen T., Mikkola T.P.J. (2004). Fatigue assessment by spectral approach for the ISSC comparative study of the hatch cover bearing pad. *Mar. Struct.*, 17, 75-90.

Łagoda T., Macha E., Pawliczek R. (2001). The influence of the mean stress on fatigue life of 10HNAP steel under random loading. *Int. J. Fatigue*, 23, 283-291.

Langlais T.E., Vogel J.H., Chase T.R. (2003). Multiaxial cycle counting for critical plane methods. *Int. J. Fatigue*, 25, 641-647.

Larsen C.E., Lutes L.D. (1991) Predicting the fatigue life of offshore structures by the single-moment spectral method. *Prob. Eng. Mech.*, 6(2), 96-108.

Lindgren G. (1970). Some properties of a normal process near a local maximum. *The Annals of Mathematical Statistics*, 41(6), 1870-1883.

Lindgren G., Rychlik I. (1987). Rain flow cycle distributions for fatigue life prediction under Gaussian load processes. *Fatigue Fract. Eng. Mater. Struct.*, 10(3), 251-260.

Lindgren G., Broberg K.B. (2003). Cycle distributions for Gaussian processes – exact and approximate results. Report 2003:26, Mathematical Statistics, Centre for Mathematical Sciences, Lund University, Sweden.

Liou H.Y., Wu W.F., Shin C.S. (1999). A modified model for the estimation of fatigue life derived from random vibration theory. *Prob. Eng. Mech.*, 14, 281-288.

Lu P., Liu X. (1997). An analytical solution of equivalent stress for structures fatigue life prediction under broad band random loading. *Mech. Struct. & Mach.*, 25(2), 139-149.

Lu P., Zhao B., Yan J. (1998). Efficient algorithm for fatigue life calculation under broad band loading based on peak approximation. *J. Eng. Mech.-ASCE*, 124(2), 233-236.

Lu P., Jiao S.J. (2000). An improved method of calculating the peak stress distribution for a broad-band random process. *Fatigue Fract. Eng. Mater. Struct.*, 23, 581-586.

Lutes L.D., Corazao M., Hu S.J., Zimmerman J. (1984). Stochastic fatigue damage accumulation. *J. Struct. Eng.-ASCE*, 110(11), 2585-2601.

Lutes L.D., Larsen C.E. (1990). Improved spectral method for variable amplitude fatigue prediction. *J. Struct. Eng.-ASCE*, 116(4), 1149-1164.

Lutes L.D., Sarkani S. (2004). Random Vibrations. Analysis of structural and mechanical Systems. Elsevier.

Madhavan Pillai T.M., Meher Prasad A. (2000). Fatigue reliability analysis in time domain for inspection strategy of fixed offshore structures. *Ocean Eng.*, 27(2), 167-186.

Madsen H.O., Krenk S, Lind N.C. (1986). Methods of structural safety. Prentice-Hall, Englewood Cliffs, New Jersey.

Matsuishi M., Endo T. (1968). Fatigue of metals subjected to varying stress. presented at Japan Society of Mechanical Engineers, Fukuoka, Japan.

Miner M.A. (1945). Cumulative damage in fatigue. *J. Appl. Mech.-Trans. ASME*, 67, A159-A164.

Mood A.M., Graybill F.A., Boes D. (1987). Introduction to the theory of statistics. McGraw-Hill, Statistical Series, 3rd edition.

Nagode M., Fajdiga M. (1998a). A general multi-modal probability density function suitable for the rainflow ranges of stationary random processes. *Int. J. Fatigue*, 20(3): 211-223.

Nagode M., Fajdiga M. (1998b). On a new method for prediction of the scatter of loading spectra. *Int. J. Fatigue*, 20(4): 271-277.

Nagode M., Klemenc J., Fajdiga M. (2001). Parametric modelling and scatter prediction of rainflow matrices. *Int. J. Fatigue*, 23, 525-532.

Ochi M.K., Ahn K. (1994). Probability distribution applicable to non-Gaussian random processes. *Prob. Eng. Mech.*, 9, 255-264.

Ochi M.K. (1998). Probability distributions of peaks and troughs of no-Gaussian random processes. *Prob. Eng. Mech.*, 13(4), 291-298.

Olagnon M. (1994). Practical computations of statistical properties of rainflow counts. *Int. J. Fatigue*, 16, 306-314.

Petrone N., Tessari A., Tovo R. (1996). Acquisition and analysis of service load histories in mountain-bikes. *Proceedings of the International Conference of Material Engineering*, Gallipoli-Lecce (Italy).

Petrucci G., Zuccarello B. (1999). On the estimation of the fatigue cycle distribution from spectral density data. *Proc. Inst. Mech. Eng. Part C-J. Mech. Eng. Sci.*, 213(8), 819-831.

Petrucci G., Di Paola M., Zuccarello B. (2000). On the Characterization of Dynamic Properties of Random Processes by Spectral Parameters. *J. Appl. Mech.-Trans. ASME*, 67(3), 519-526.

Petrucci G., Zuccarello B. (2004). Fatigue life prediction under wide-band random loading. *Fatigue Fract. Eng. Mater. Struct.*, 27(12), 1183-1195.

Pitoiset X., Preumont A., Kernilis A. (1998). Tools for a multiaxial fatigue analysis of structures submitted to random vibrations. *Proceedings of the European Conference on Spacecraft Structures, Materials and Mechanical Testing*, Braunschweig, Germany.

Pitoiset X., Preumont A. (2000). Spectral methods for multiaxial random fatigue analysis of metallic structures. *Int. J. Fatigue*, 22(7), 541-550.

Pitoiset X. (2001). Méthodes spectrales pour une analyse en fatigue des structures métalliques sous chargements aléatoires multiaxiaux. PhD thesis, Faculté de Science Appliquées, Université Libre de Bruxelles, Bruxelles (France) (in French).

Pitoiset X., Rychlik I., Preumont A. (2001). Spectral methods to estimate local multiaxial fatigue failure for structures undergoing random vibrations. *Fatigue Fract. Eng. Mater. Struct.*, 24, 715-727.

Rejman A., Rychlik I. (1993). Fatigue life distribution with linear and nonlinear damage rules. Research Report, 1993:3, Dept. Math. Statist., Lund University (Sweden).

Robson J.D., Dodds C.J. (1975/76). Stochastic road inputs and vehicle response. *Vehicle Syst. Dyn.*, 5, 1-13.

Rouillard V., Sek M.A. (2000). Monitoring and simulating non-stationary vibrations for package optimisation. *Packag. Technol. Sci.*, 13, 149-156.

Rouillard V. (2002). Remote monitoring of vehicle shock and vibrations. *Packag. Technol. Sci.*, 15, 83-92.

Rychlik I. (1987). A new definition of the rain-flow cycle counting method. *Int. J. Fatigue*, 9(2), 119-121.

Rychlik I. (1989). Simple approximations of the Rain-Flow-Cycle distribution for discretized loads. *Prob. Eng. Mech.*, 4(1), 40-48.

Rychlik I. (1993a). On the 'narrow-band' approximation for expected fatigue damage. *Prob. Eng. Mech.*, 8, 1-4.

Rychlik I. (1993b). Note on cycle counts in irregular loads. *Fatigue Fract. Eng. Mater. Struct.*, 16(4), 377-390.

Rychlik I. (1993c). Characterisation of random fatigue loads. In: *Stochastic approach to fatigue*, CISM Courses n. 334, Ed. Sobczyk K.Springer-Verlag.

Rychlik I. (1996). Fatigue and stochastic loads. *Scan. J. Stat.*, 23, 387-404.

Rychlik I., Johannesson P., Leadbetter M.R. (1997). Modelling and statistical analysis of ocean-wave data using transformed Gaussian processes. *Mar. Struct.*, 10, 13-47.

Sakai S. Okamura H. (1995). On the distribution of rainflow range for Gaussian random processes with bimodal PSD. *JSME Int. Journal*, Series A, 38(4), 440-445.

Sarkani S., Kihl D.P., Beach J.E. (1994). Fatigue of welded joints under narrowband non-Gaussian loadings. *Prob. Eng. Mech.*, 9, 179-190.

Sarkani S., Michaelov G., Kihl D.P., Beach J.E. (1996). Fatigue of welded joints under wideband loadings. *Prob. Eng. Mech.*, 11, 221-227.

Schütz D., Klätschke H., Steinhilber H., Heuler P., Schütz W. (1990). Standardized load sequences for car wheel suspension components (CARLOS). LBF Report n. FB-191.

Schütz W. (1996). A history of fatigue. *Eng. Fract. Mech.*, 54(2), 263-300.

Shinozuka M., Jan C.M. (1972). Digital simulation of random processes and its applications. *J. Sound Vibr.*, 25(1), 111-128.

Shinozuka M., Deodatis G. (1991). Simulation of stochastic processes by spectral representation. *Appl. Mech. Rev.*, 44(4), 191-204.

Siddiqui N.A., Ahmad S. (2001). Fatigue and fracture reliability of TLP tethers under random loading. *Mar. Struct.*, 14, 331-352.

Sjöström S. (1961). On random load analysis. *Trans. of the Royal Institute of Technology*, Stockholm, n. 161.

Sonsino C.M. (2007a). Course of SN-curves especially in the high-cycle fatigue regime with regard to component design and safety. *Int. J. Fatigue*, 29(12), 2246–2258.

Sonsino C.M. (2007b). Fatigue testing under variable amplitude loading. *Int. J. Fatigue*, 29(6), 1080-1089.

Stichel S., Knothe K. (1998). Fatigue life prediction for a S-train bogie. *Vehicle System Dynamics Supplement*, 28, 390-403.

Sun L. (2001). Computer simulation and field measurement of dynamic pavement loading. *Math. Comput. Simul.*, 56, 297-313.

Sun. L., Kennedy T.W. (2002). Spectral analysis and parametric study of stochastic pavement loads. *J. Eng. Mech.-ASCE*, 128(3), 318-327.

Susmel L., Tovo R., Benasciutti D. (2009). A novel engineering method based on the critical plane concept to estimate lifetime of weldments subjected to variable amplitude multiaxial fatigue loading. *Fatigue Fract Eng. Mater. Struct.*, 32(5), 441-459.

Svensson T. (1997). Prediction uncertainties at variable amplitude fatigue. *Int. J. Fatigue*, 19(Suppl.1), S295-S302.

Tovo R. (2000). A damaged based evaluation of probability density distribution for rain-flow ranges from random processes. *Int. J. Fatigue*, 22, 425-429.

Tovo R. (2001). On the fatigue reliability evaluation of structural components under service loading. *Int. J. Fatigue*, 23, 587-598.

Tovo R. (2002). Cycle distribution and fatigue damage under broad-band random loading. *Int. J. Fatigue*, 24(11), 1137-1147.

Tunna J.M. (1985). Random fatigue: theory and experiment. *Proc. Inst. Mech. Eng. Part C-J. Mech. Eng. Sci.*, 199(C3), 249-257.

Tunna J.M. (1986). Fatigue life prediction for Gaussian random loads at the design stage. *Fatigue Fract. Eng. Mater. Struct.*, 9(3), 169-184.

Vanmarcke E.H. (1972). Properties of spectral moments with applications to random vibration. *J. Eng. Mech.-ASCE*, 98, 425-446.

Winterstein S.R. (1985). Non-normal responses and fatigue damage. *J. Eng. Mech.-ASCE*, 111(10), 1291-1295.

Winterstein S.R. (1988). Nonlinear vibration models for extremes and fatigue. *J. Eng. Mech.-ASCE*, 114(10), 1772-1790.

Winterstein S.R., Ude T.C., Kleiven G. (1994). Springing and slow drift responses: predicted extremes and fatigue vs. simulation. *Proceedings of the 7th International behaviour of Offshore structures BOSS*, Cambridge, 3, 1-15.

Wirsching P.H., Sheata A.M. (1977). Fatigue under wide band random stresses using the rain-flow method. *J. Eng. Mater. Technol.-Trans. ASME*, 99, 205-211.

Wirsching P.H., Light C.L. (1980). Fatigue under wide band random stresses. *J. Struct. Division ASCE*, 106 (7), 1593-1607.

Withey P.A. (1997). Fatigue failure of the de Havilland Comet I. *Eng. Fail. Anal.*, 4(2), 147-154.

Wu W.F., Huang T.H. (1993). Prediction of fatigue damage and fatigue life under random loading. *Int. J. Pressure Vessels Pip.*, 53, 273-298.

Wu W.F., Liou H.Y., Tse H.C. (1997). Estimation of fatigue damage and fatigue life of components under random loading. *Int. J. Pressure Vessels Pip.*, 72, 243-249.

Yu L., Das P.K., Barltrop D.P. (2004). A new look at the effect of bandwidth and non-normality on fatigue damage. *Fatigue Fract. Eng. Mater. Struct.*, 27(1), 51-58.

Zhao W., Baker M.J. (1992). On the probability density function of rainflow stress range for stationary Gaussian processes. *Int. J. Fatigue*, 14(2), 121-135.

LIST OF SYMBOLS

DK	Dirlik
FC	Fu-Cebon
G	Gaussian
nG	non-Gaussian
MFC	modified Fu-Cebon
JM	Jiao-Moan
NB	narrow-band approximation
PV	peak approximation
PZ	Petrucci-Zuccarello
RC	range counting
RFC	rainflow counting
TB	Tovo-Benasciutti
SM	single-moment
SO	Sakai-Okamura
ZB	Zhao-Baker
α_2, α_X	irregularity index for process $X(t)$
$\alpha_{\dot{X}}$	irregularity index for process $\dot{X}(t)$
α_m	set of bandwidth parameters for process $X(t)$
β_m	set of bandwidth parameters for derivative process $\dot{X}(t)$
γ_3, γ_4	skewness, kurtosis
$\Gamma(\cdot)$	gamma function
$\delta(\cdot)$	Dirac delta function
ε	spectral width parameter
$\zeta(x)$	road irregularity (spatial random process)
η	damage ratio (i.e. non-Gaussian to Gaussian damage)
κ_1, κ_2, κ_3	cumulants
K	scale factor in the Hermite model
λ_m	m-th spectral moment
$\lambda_{m,1}$, $\lambda_{m,2}$	m-th spectral moment for $X_1(t)$ and $X_2(t)$ components

$\mu_X(t)$, $\mu_Z(t)$	mean value of process $X(t)$ and $Z(t)$
$\mu(u,v)$	count intensity
$\mu_T(u,v)$	expected count distribution
$\mu_T(\cdot)$	level upcrossing spectrum in time T
$\nu(\cdot)$	level upcrossing intensity
ν_0	mean upcrossing intensity (i.e. mean upcrossings/sec)
$\nu_{0,1}$, $\nu_{0,2}$	mean upcrossing intensity of $X_1(t)$ and $X_2(t)$
ν_a	cycle intensity (i.e. counted cycles/sec)
ν_p	peak intensity (i.e. peaks/sec)
ξ	damping coefficient
$\Phi(\cdot)$	standard normal distribution function
ρ	damage correction factor
σ_X^2	variance of process $X(t)$
σ_{xx}, σ_{yy}	stress components along x- and y-direction
τ_{xy}	shear stress in the xy-plane
τ_{eq}	equivalent shear stress
ω	angular frequency
ω_1, ω_2	low and high central frequency
ω_D, ω_N	dominant and structural angular frequencies
b, b_{app}	weighting coefficient
B	area ratio
c	frequency ratio (single-block)
C	fatigue strength for the S-N curve
$d(\cdot)$	damage function (i.e. damage for one cycle)
D_{cr}	critical value for fatigue damage
$\overline{D}_0$	expected damage intensity for the reference model
$\overline{D}^{G}$, $\overline{D}^{nG}$	Gaussian and non-Gaussian damage intensity
$D(T)$	fatigue damage in time period T
$D^+(T)$	upper bound of fatigue damage
$\overline{D}(T)$	expected fatigue damage in time period T
$\overline{D}(1)$, $\overline{D}$	expected fatigue damage in unit time (damage intensity).
$E[\cdot]$	stochastic mean (expectation operator)
f	frequency
$\mathbf{F}$, $\hat{\mathbf{F}}$	observed min-max, max-min matrix
$\mathbf{F}'$	observed from-to matrix
$\mathbf{F}^{rfc}$	observed rainflow matrix
$F(s)$	fatigue loading (or cumulative) spectrum
$F_1(t)$, $F_2(t)$	forces in the car-quarter model
$G(\cdot)$, $g(\cdot)$	direct and inverse transformation

h_3, h_4, $\tilde{h}_3$, $\tilde{h}_4$	parameters of the Hermite model
$h(u,v)$, $H(u,v)$	probability density and cumulative distribution of counted cycles as a function of maximum and minimum
$H_1(\omega)$, $H_2(\omega)$	harmonic transfer function for the forces $F_1(t)$ and $F_2(t)$
IF	irregularity factor
$\mathbf{I}(\cdot)$	indicator function
k	slope of the S-N curve
m	mean value of a counted cycle
M_k, m_k	maximum and minimum of a counted cycle (e.g. m_k^{rc} for the range count, m_k^{rfc} for the rainflow count)
N	number of cycles to failure at constant amplitude loading
$\overline{N}$	expected number of cycles counted in time period T
$\overline{N}_{0,1}$, $\overline{N}_{0,2}$	Expected number of mean upcrossings for $X_1(t)$ and $X_2(t)$
$\overline{N}_S$, $\overline{N}_L$	expected number of small and large cycles in time T
$N(T)$	number of cycles counted in time period T
$N_T(u,v)$	count distribution
$p_a(s)$	amplitude probability density
$p_{a,m}(s,m)$	amplitude and mean probability density
$p_{s_1}(s)$, $p_{s_2}(s)$	probability density of amplitudes s_1 and s_2
$p_{s_S}(s)$, $p_{s_L}(s)$	probability density of small and large cycles
$p_p(u)$, $P_p(u)$	probability density and cumulative distribution of peaks
$p_v(v)$, $P_v(v)$	probability density and cumulative distribution of valleys
$\mathbf{P}$, $\hat{\mathbf{P}}$	min-max, max-min transition matrix
$\mathbf{Q}$, $\hat{\mathbf{Q}}$	expected min-max, max-min matrix
$\mathbf{Q}^{rfc}$	expected rainflow matrix
$P(t)$	envelope process of $X(t)$
q_X, $q_{\dot{X}}$	Vanmarcke's bandwidth parameter for $X(t)$, $\dot{X}(t)$
$Q(t)$	amplitude process of $X(t)$
r	range of a counted cycle
R	frequency ratio
$R(t)$	Cramer-Leadbetter envelope process
$r(\cdot)$, $R(\cdot)$	Rayleigh probability density and cumulative distribution
s	amplitude of a counted cycle
s_1, s_2	amplitudes associated to $X_1(t)$ and $X_2(t)$ processes
s_S, s_L	amplitude of small and large cycles in a bimodal process
$S_X(\omega)$	two-sided spectral density of process $X(t)$
S_u	ultimate tensile stress
t	time

T	time period
T_D, T_N	dominant wave period, dominant structural period
u, v	peak, valley
$W_\zeta(n)$	one-sided spectral density of process $\zeta(x)$
$W_X(\omega)$	one-sided spectral density of process $X(t)$
$W_1(\omega)$, $W_2(\omega)$	one-sided spectral density of forces $F_1(t)$ and $F_2(t)$
$(x_p, x_v), (z_p, z_v)$	Gaussian and non-Gaussian cycle (peak and valley)
$X(t)$, $Z(t)$	random processes (e.g. Gaussian and non-Gaussian)
$X_1(t)$, $X_2(t)$	low- and high-frequency narrow-band component
$\bar{z}(t)$	non-stationary non-Gaussian loading
Z	normalised amplitude

Appendix A

PROPERTIES OF BANDWIDTH PARAMETERS

A.1. THE GENERALISED SCHWARTZ'S INEQUALITY

Let λ_m be the spectral moments associated to spectral density $S_X(\omega)$ of process $X(t)$; further, let α_m and β_m be the set of bandwidth parameters of process $X(t)$ and its derivative $\dot{X}(t)$, see Chapter 3. This section, by using some general arguments involving α_m and β_m parameters, will show how $\alpha_1 \geq \alpha_2$ and $\beta_1 \geq \beta_2$, and further that $\alpha_{0.75} \geq \alpha_1$.

The Schwartz's inequality for spectral moments λ_m writes [Lutes and Sarkani 2004]:

$$0 \leq \lambda_m^2 \leq \lambda_0 \, \lambda_{2m} \tag{A.1}$$

which proves that $0 \leq \alpha_m \leq 1$; consequently, $0 \leq \alpha_1 \leq 1$, $0 \leq \alpha_2 \leq 1$ and $0 \leq \alpha_{0.75} \leq 1$.

Let us now introduce the spectral density $U_X^n(\omega)$, defined from $S_X(\omega)$ through integer n:

$$U_X^n(\omega) = \omega^n \, S_X(\omega) \tag{A.2}$$

For $n = 1, 2$, $U_X^n(\omega)$ represents the spectral density of process $X(t)$ and its derivative $\dot{X}(t)$.

Let $\tilde{\lambda}_m$ be the spectral moments of $U_X^n(\omega)$; obviously $\tilde{\lambda}_m = \lambda_{m+n}$. The Schwartz's inequality, Eq. (A.1), applied to spectral moments $\tilde{\lambda}_m$ writes:

$$\tilde{\lambda}_m^2 \leq \tilde{\lambda}_0 \, \tilde{\lambda}_{2m} \tag{A.3}$$

and it specialises in terms of moments λ_m as:

$$0 \leq \lambda_{n+m}^2 \leq \lambda_n \, \lambda_{2m+n} \tag{A.4}$$

We shall refer to Eq. (A.4) as the "generalised Schwartz's inequality". By taking $n = 2$, equation above shows that $0 \le \beta_m \le 1$.

A.1.1. Lemma 1 - Proof that $\alpha_m \ge \alpha_{2m}$

For process $X(t)$, we have by definition:

$$\alpha_m = \frac{\lambda_m}{\sqrt{\lambda_0 \lambda_{2m}}} \qquad \alpha_{2m} = \frac{\lambda_{2m}}{\sqrt{\lambda_0 \lambda_{4m}}} \qquad (A.5)$$

thus in order to prove that $\alpha_m \ge \alpha_{2m}$, we must show that $\lambda_{2m}^3 \le \lambda_m^2 \lambda_{4m}$. For a given m value, the generalised Schwartz's inequality writes as in Eq. (A.4); by taking $n = m$, it writes:

$$\lambda_{m+m}^2 \le \lambda_m \lambda_{2m+m} \qquad \rightarrow \qquad \lambda_{2m}^2 \le \lambda_m \lambda_{3m} \qquad (A.6)$$

and taking a new index $\tilde{n} = n + m = m + m$, it writes:

$$\lambda_{\tilde{n}+m}^2 \le \lambda_{\tilde{n}} \lambda_{2m+\tilde{n}} \qquad \rightarrow \qquad \lambda_{3m}^2 \le \lambda_{2m} \lambda_{4m} \qquad (A.7)$$

Squaring Eq. (A.6) and using Eq. (A.7), we have:

$$\lambda_{2m}^4 \le \lambda_m^2 \lambda_{3m}^2 \le \lambda_m^2 \left(\lambda_{2m} \lambda_{4m} \right) \qquad \rightarrow \qquad \lambda_{2m}^3 \le \lambda_m^2 \lambda_{4m} \qquad (A.8)$$

which finishes the proof □.

A.1.2. Lemma 2 – Proof that $\beta_m \ge \beta_{2m}$

For process $\dot{X}(t)$, we have by definition:

$$\beta_m = \sqrt{\frac{\lambda_{m+2}^2}{\lambda_2 \lambda_{2m+2}}} \quad , \quad \beta_{2m} = \sqrt{\frac{\lambda_{2m+2}^2}{\lambda_2 \lambda_{4m+2}}} \qquad (A.9)$$

thus proving that $\beta_m \ge \beta_{2m}$ is equivalent to show that $\lambda_{2m+2}^3 \le \lambda_{m+2}^2 \lambda_{4m+2}$. For a given m value, the generalised Schwartz's inequality writes as in Eq. (A.4). By taking $n = m + 2$, it writes:

$$\lambda_{(m+2)+m}^2 \le \lambda_{m+2} \lambda_{2m+(m+2)} \qquad \rightarrow \qquad \lambda_{2m+2}^2 \le \lambda_{m+2} \lambda_{3m+2} \qquad (A.10)$$

and taking now a new index $\tilde{n} = n + m = (m + 2) + m$, it writes:

$$\lambda_{\tilde{n}+m}^2 \le \lambda_{\tilde{n}} \lambda_{2m+\tilde{n}} \qquad \rightarrow \qquad \lambda_{3m+2}^2 \le \lambda_{2m+2} \lambda_{4m+2} \qquad (A.11)$$

By combining previous results, we get:

$$\lambda^4_{2m+2} \leq \lambda^2_{m+2} \, \lambda^2_{3m+2} \leq \lambda^2_{m+2} \left(\lambda_{2m+2} \, \lambda_{4m+2} \right) \quad \rightarrow \quad \lambda^3_{2m+2} \leq \lambda^2_{m+2} \, \lambda_{4m+2}$$

(A.12)

which concludes the proof □.

As a consequence of setting $m = 1$, previous Lemmas show that $\alpha_1 \geq \alpha_2$ and $\beta_1 \geq \beta_2$.

A.1.3. Lemma 3 – Proof that $\alpha_{0.75} \geq \alpha_1$

Index m and n can take also non-integer values. By definition we have:

$$\alpha_{0.75} = \frac{\lambda_{0.75}}{\sqrt{\lambda_0 \, \lambda_{1.5}}} \qquad \alpha_1 = \frac{\lambda_2}{\sqrt{\lambda_0 \, \lambda_2}}$$

(A.13)

thus proving that $\alpha_{0.75} \geq \alpha_1$ is equivalent to show that $\lambda^3_1 \, \lambda_{1.5} \leq \lambda_{0.75} \, \lambda_2$. From Eq. (A.1), we can write the following list of Schwartz's inequalities:

$$\lambda^2_1 \leq \lambda_{0.75} \, \lambda_{1.25} \qquad (m = 0.25 \, , \, n = 0.75)$$

(A.14)

$$\lambda^2_{1.25} \leq \lambda_1 \, \lambda_{1.5} \qquad (m = 0.25 \, , \, n = 1)$$

(A.15)

$$\lambda^2_{1.5} \leq \lambda_1 \, \lambda_2 \qquad (m = 0.5 \, , \, n = 1)$$

(A.16)

By taking the square of Eq. (A.14) and using Eq. (A.15) we obtain:

$$\lambda^4_1 \leq \lambda^2_{0.75} \, \lambda^2_{1.25} \leq \lambda^2_{0.75} \left(\lambda_1 \, \lambda_{1.5} \right)$$

(A.17)

that simplifies into:

$$\lambda^3_1 \leq \lambda^2_{0.75} \, \lambda_{1.5}$$

(A.18)

Multiplying now for $\lambda_{1.5}$ and substituting Eq. (A.16) we have:

$$\lambda^3_1 \, \lambda_{1.5} \leq \lambda^2_{0.75} \, \lambda^2_{1.5} \leq \lambda^2_{0.75} \, \lambda_1 \, \lambda_2$$

(A.19)

that can be simplified into:

$$\lambda^2_1 \, \lambda_{1.5} \leq \lambda^2_{0.75} \, \lambda_2$$

(A.20)

thus finishing the proof □.

203

Appendix B

THE APPROXIMATE b WEIGHTING FACTOR

B.1. PRELIMINARY CONSIDERATIONS

In a Gaussian random process $X(t)$, the rainflow damage rate for TB method is:

$$\overline{D}_{\text{RFC}} = b \, \overline{D}_{\text{NB}} + (1-b) \, \overline{D}_{\text{RC}} \tag{B.1}$$

being $\overline{D}_{\text{NB}}$ and $\overline{D}_{\text{RC}}$ the damage intensities according to the narrow-band approximation and the range counting method, respectively. Coefficient b is a dimensionless weighting factor assumed dependent on $S_X(\omega)$, the spectral density of process $X(t)$. However, no theoretical information is available about this dependence, so we must use approximations.

This section aims to develop a suitable approximate expression for b coefficient, on the basis of results from numerical simulations presented in Chapter 4.

From a general theoretical point of view, b should depend on spectral density $S_X(\omega)$ through the entire set of spectral moments λ_m.

However, the need of finding a quite simple expression for b suggests we adopt some simplifying hypotheses. The main assumption is that b depends only on α_1 and α_2 bandwidth parameters as a two-variable function, i.e. $b = b(\alpha_1, \alpha_2)$. For example, an approximate formula already exists [Tovo 2002]:

$$b_{\text{app}}^{\text{Tov}} = \min\left(\frac{\alpha_1 - \alpha_2}{1 - \alpha_1}, 1 \right) \tag{B.2}$$

which is shown in Figure B.1.

The b index defined in Eq. (B.2) is not continuous and numerical simulations pointed out that it may give inaccurate results when α_1 and α_2 are quite different (e.g. when α_1 approaches unity and $\Delta = \alpha_1 - \alpha_2$ increases). A typical example of such a situation may be encountered with a process representing the response of a linear oscillator driven by white noise, for which $\alpha_1 \approx 1$ and $\alpha_2 = 0$ (irregular process).

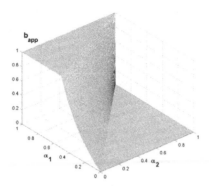

Figure B.1: The approximate b factor as given by Eq. (B.2).

Our aim is to improve expression above still using a simple mathematical expression as a continuous function of some spectral parameters, calibrated through a best fitting procedure over all simulation results. Since all spectral densities considered in simulations differ only in respect to α_1 and α_2 bandwidth parameters, the assumption of a two-variable function $b = b(\alpha_1, \alpha_2)$ is still maintained.

B.2. THE POLYNOMIAL MODEL

A critical analysis of the b values provided by numerical simulations (see Chapter 4) highlights two important features of the b coefficient:

1) for any α_2 value, b is observed to converge to zero when $\alpha_1 \to \alpha_2$;
2) for any α_2 value, b is observed to converge to one when $\alpha_1 \to 1$.

The simplest mathematical model for b seems to be a two-variable function in the form of a quadratic polynomial in α_1:

$$b(\alpha_1, \alpha_2) = m(\alpha_2)\alpha_1^2 + n(\alpha_2)\alpha_1 + p(\alpha_2) \qquad (B.3)$$

whose coefficients are supposed to be dependent on α_2 alone; all polynomial coefficients are completely specified by introducing an additional condition on b.

Based on results from simulation results, we observe that a possible condition should be to impose a prescribed slope for $\alpha_1 = \alpha_2$ for the polynomial; this slope is assumed an exponential function of α_2, as:

$$t(\alpha_2) = A \, e^{B \alpha_2} \qquad (B.4)$$

205

Expressing all condition gives the following result:

$$\begin{cases} b(\alpha_1 = \alpha_2) = 0 & \rightarrow & m\,\alpha_2^2 + n\,\alpha_2 + p = 0 \\ b(1, \alpha_2) = 1 & \rightarrow & m + n + p = 1 \\ \partial_{\alpha_1} b(\alpha_1 = \alpha_2) = A\,e^{B\,\alpha_2} & \rightarrow & 2\,m\,\alpha_2 + n = A\,e^{B\,\alpha_2} \end{cases} \tag{B.5}$$

where ∂_{α_1} denotes the derivative respect to α_1. Equation above is a linear system in variables m, n and p (i.e. the coefficients of the polynomial), with solution:

$$\begin{cases} m = \dfrac{A\,e^{B\,\alpha_2}(\alpha_2 - 1) + 1}{(\alpha_2 - 1)^2} \\[3mm] n = \dfrac{A\,e^{B\,\alpha_2}(1 - \alpha_2^2) - 2\alpha_2}{(\alpha_2 - 1)^2} \\[3mm] p = \alpha_2\,\dfrac{A\,e^{B\,\alpha_2}(\alpha_2 - 1) + \alpha_2}{(\alpha_2 - 1)^2} \end{cases} \tag{B.6}$$

Substituting these solutions into the polynomial given in Eq. (B.3) and then simplifying, gives the following final form for the $b(\alpha_1, \alpha_2)$ coefficient:

$$b_{app} = \frac{(\alpha_1 - \alpha_2)[A\,(1 + \alpha_1\alpha_2 - (\alpha_1 + \alpha_2))\,e^{B\,\alpha_2} + (\alpha_1 - \alpha_2)]}{(\alpha_2 - 1)^2} \tag{B.7}$$

Expression depends on only two parameters A and B; a best-fitting procedure on results obtained in Chapter 4 gives the values $A = 1.112$ and $B = 2.11$; the final version of b_{app} is shown in Figure B.2.

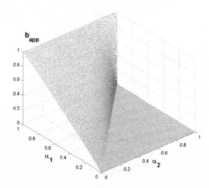

Figure B.2: The approximate b factor as given by Eq. (B.7), in which $A = 1.112$ and $B = 2.11$.